RIBEIRO · PHYTOPHTHORA

A SOURCE BOOK
OF THE GENUS *PHYTOPHTHORA*

by

OLAF K. RIBEIRO

Illustrated

1978 · J. CRAMER

In der A.R. Gantner Verlag Kommanditgesellschaft

FL - 9490 VADUZ

Author's Address:

Department of Plant Pathology,
University of California,
Riverside, California 92521,
U.S.A.

©1978 A.R. Gantner Verlag K.-G., FL-9490 Vaduz
Printed in Germany
by Strauss & Cramer GmbH, D-6945 Hirschberg II
ISBN 3-7682-1200-9

Contents

Tables

Figures

Preface

The genus *Phytophthora* first came into prominence with the discovery that the disease devastating potato crops in Ireland in the late 1800's, was caused by *Phytophthora infestans*. Since that time, *Phytophthora* species have been identified as the causal agents responsible for severe losses in several economically important plants. Diseases such as black rot of cocoa pods; avocado root rot; Jarrah dieback of eucalyptus; black stripe of rubber; foot rot or gummosis of citrus; red stele of strawberries; blight of potatoes and tomatoes; decay of stems and roots of several forage, vegetable and ornamental plants; and root rot of deciduous fruit trees, indicate the vast range of hosts parasitized by this genus.

My personal involvement with *Phytophthora* began as a graduate student working on the genetics and physiology of the fungus, with Prof. M.E. Gallegly at West Virginia University. This was an exciting period in *Phytophthora* research; the sexual stage of *P. infestans* had just been discovered in Mexico by J.A. Galindo and M.E. Gallegly; dominant genes for resistance to potato blight were being identified; the nutritional requirements for growth and reproduction of the fungus were being painstakingly elucidated by Profs. V.G. Lilly and H.L. Barnett; sterols were found to be essential for reproduction; a key to the species of *Phytophthora* had just been published; and a selective antibiotic medium for the direct isolation of *Phytophthora* from soil was developed by Prof. P.H. Tsao at the University of California, Riverside. These findings gave impetus to world-wide research on various species of the genus. By the mid-1970's, research at the University of California, Riverside, indicated that this fungus has many unique features. This was also the period of my involvement with *Phytophthora* investigations at Riverside, and hence gave me an opportunity to personally share in some of the excitement generated as each new discovery was made on various aspects of the organism.

We now know that the fungus is diploid in the vegetative stage, and from Prof. S. Bartnicki-Garcia's work, we discover that *Phytophthora* has an interesting cell wall structure. This demonstration that *Phytophthora* has cellulose rather than chitin as the main structural component of the cell wall, makes this fungus unique among the Phycomycetes. Many other exciting findings in the areas of ultrastructure, biochemistry, and physiology, are described in the text.

During my association with the genus *Phytophthora*, it became apparent that there existed a need for a central source of information on various aspects of the genus which could be readily accessible. Many of the techniques for routine production of sporangia or oospores, isolations, pathogenicity tests, identification, etc. are scattered through numerous scientific journals making information retrieval a long and tedious task, particularly in areas where library facilities are limited.

It is hoped then, that this compilation will fill the need for a ready source of reference for the genus *Phytophthora*. Many of the techniques described have been used successfully in our research. Some of these techniques have not been published before. Sufficient alternatives are given whenever possible to help investigators in situations where facilities or equipment are limiting factors. This book is not intended to be a complete source of reference for all the literature published on the genus *Phytophthora*, but rather should be viewed as synopsis of techniques and information which provide a quick source of reference to many aspects of the fungus. Sufficient key references have been added wherever possible to permit detailed studies when desirable.

As is inevitable with a compilation of this kind, some key sources of reference and/or techniques have been inadvertently overlooked. This is not intentional, and I would be grateful if such omissions are brought to my attention for inclusion in future editions. It is my hope that this book will prove to be a useful addition to the library of investigators actively engaged in research on *Phytophthora*, and to those who wish to know more about this unique genus of plant pathogens. Much still needs to be known about *Phytophthora*. We still do not know the stimulus involved in sexual reproduction, or the mechanism(s) of dormancy and germination of oospores. The rapid appearance of new complex races of *P. infestans* in nature are still a matter of conjecture, and control measures that are in harmony with the environment still elude us.

The task of compiling much of the information contained in this book has been greatly facilitated by authors who so readily gave permission to abstract from their published works; to them I owe a debt of gratitude. In particular, I wish to acknowledge the help of Profs. G.A. Zentmyer and D.C. Erwin for use of their laboratory facilities, and for their expert advice on the genus *Phytophthora*; and to Profs. D.E. Hemmes, N.T. Keen, S. Bartnicki-Garcia, and P.H. Tsao, for many valuable and interesting discussions on specific aspects of *Phytophthora*. Much useful information was also provided by Profs. F.J. Newhook and A.F. Schmitthenner for which I am grateful. I also wish to express my appreciation to Ms. Wendy Reid for her considerable talents in drawing up the chart of *Phytophthora* species.

Acknowledgements

The author wishes to express his appreciation to the following
publishers for permission to use material from copyrighted publications:

American Phytopathological Society, St. Paul, Minn. U.S.A.
(Phytopathology).

Cambridge University Press, New York. (Transactions of the British Mycological Society and Journal of General Microbiology.)

Commonwealth Mycological Institute, Kew, England. (Mycological
Papers Nos. 92 and 122.)

Longman Group Ltd. Essex, England.

Controller of Her Majesty's Stationery Office, London, England.
(Plant Pathology.)

National Research Council of Canada (Canadian Journal of Botany,
and Canadian Journal of Microbiology).

The New York Botanical Garden, Bronx, N.Y. (Mycologia.)

Springer-Verlag, New York Inc. (Archives of Microbiology.)

John Wiley & Sons, Inc., New York.

INTRODUCTION

HISTORY

The genus PHYTOPHTHORA was erected by Anton de bary, in 1876.
(Greek: PHYTON & PHTHEIRO = 'Plant Destroyer'.)

All described species are pathogens, attacking the leaves, flowers, stems, buds, fruits, crowns, and roots of host plants.
The genus gained prominence with the devastation wreaked by
Phytophthora infestans on potatoes. Species within the genus are
now known to the important economic pathogens of avocado, cacao,
citrus, coconut, forest trees, ornamentals, rubber, vegetables,
etc.

CHARACTERISTICS OF THE GENUS PHYTOPHTHORA

MYCELIUM (Gr. *mykes*, a fungus: *-lium*, after epithelium).

Hyaline, branching, coenocytic, becoming septate with age. Variable in diameter (5-8 μM), sometimes swollen, nodose, or tuber-
culate. Diameter dependent on physical and chemical nature of
the medium, and whether the mycelium is on the surface, aerial,
submerged, or within host cells. Hyphae branch at right angles
and are often constricted at the base. Branching is influenced
by above-mentioned factors.

ASEXUAL REPRODUCTION.

By development of sporangia and/or chlamydospores, depending on the species.

SPORANGIA. (Gr. *spora*, a seed: *angeion*, a vessel).

Sporangia are produced on sporangiophores which differ slightly or not at all from vegetative hyphae. In some species the sporangiophore resumes growth through the base of the evacuated sporangium; the new sporangia proliferate within the walls of the empty one or the sporangiophore may grow out through the exit pore of the previous sporangium and form the next sporangium some distance from the last. Sporangia are variable in size and shape, usually subspherical, ovoid, limoniform, pyriform, or obpyriform; hyaline to light yellow.

Sporangia germinate directly by one to several germ tubes or indirectly by division of the contents within the spore into an indefinite number of zoospores.

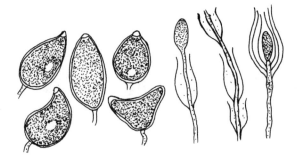

PAPILLA. (Lat. *papilla*, a nipple, a bud).

Mucilaginous area of the inner sporangium wall which when fresh will imbibe water and finally dissolve to allow emission of zoospores. The papilla has a different refractive index from the rest of the sporangium.

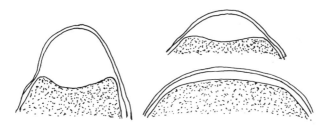

ZOOSPORES.

Zoospores emerge in amoeboid fashion through an exit pore smaller than the diameter of the zoospores. The exit pore is formed by the dissolution of the apical papillum. Zoospores are reniform in shape with two flagellae attached near the middle of the concave or flattened side. Zoospores encyst after swimming for a few minutes to several hours; they germinate by formation of germ tubes.

CHLAMYDOSPORES. (Gr. *chlamys*, *chlamd-*, a cloak; *sporā*, a seed).

Spherical to ovoid; non-papillate; hyaline to deep brown; thin to thick walled; terminal or inter-calary. Distinguished from hyphal swellings by the presence of a septum and from oospores by its strong birefringence and lack of a separate (oogonial) wall.

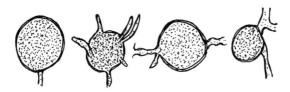

SEXUAL REPRODUCTION.

Sexual reproduction is by antheridial and oogonial interaction resulting in the formation of an oospore; presumably by fusion of gametangial nuclei.

OOGONIUM. (Gr. *oōn-*, an egg; *gon-*, race).

Spherical to pyriform, smooth, hyaline to yellowish. Delimited
from hypha by a septum. Enclosed is an oosphere which after fertilization develops into a single smooth, spherical, hyaline to
yellowish OOSPORE (*oōn*, an egg; *sporā*, a seed), nearly filling
the interior of the oogonium.

ANTHERIDIUM. (Gr. *antheros*, flowering).

Usually produced as a multinucleate swollen hyphal tip cut off
by a septum. Attaches firmly to the oogonium at an early stage
of development. Two types are known; amphigynous and paragynous.

Paragynous. (Gr. *para*, beside; *gyn-*,, female).

Attached to one side of the oogonium either near the oogonial
stalk, or elsewhere on the oogonium.

Amphigynous. (Gr. *amphi*, about, on both sides; *gyn-*, female).

Tightly surrounds the stalk of the oogonium by which it was
penetrated. Remains attached permanently.

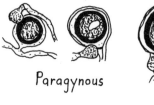 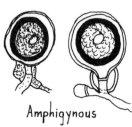

Paragynous

Amphigynous

HYPHAL SWELLINGS.

Characteristic of certain species e.g. *P. cinnamomi*, *P. crypto-gea*, and *P. megasperma*. May be single, crowded, spherical, intercalary, or terminal. Not cut off by a septum c/f chlamydospores. Hyphal swellings are sometimes formed in some species growing in abnormal conditions e.g. low temperatures or unsuitable growth medium.

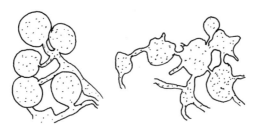

References:

BLACKWELL, E. 1949. Terminology in *Phytophthora*. Mycol. Papers 30: 1-25.

TUCKER, C.M. 1931. Taxonomy of the genus *Phytophthora* de Bary. Missouri Agr. Expt. Sta. Bull. 153: 207 pp.

RIBEIRO, O.K. 1976. Dept. of Plant Pathology, University of California, Riverside, Ca.

Section 1:

Hosts and Disease Symptoms

Table 1: Host Specialization of *Phytophthora* Species

General	Partially Specialized	Specialized	Extremely Specialized
Numerous Hosts	Several Hosts	2-3 Hosts	One Host
P. cactorum	P. cambivora	P. arecae	P. castaneae
P. cinnamomi	P. capsici	P. bahamensis	P. cyperi
P. cryptogea	P. citricola	P. boehmeriae	P. cyperi-bulbosi
P. drechsleri	P. citrophthora	P. botryosa	P. inflata
P. megasperma var. megasperma	P. erythroseptica	P. carica (fici)	P. lepironiae
P. nicotianae var. nicotianae	P. fragariae	P. infestans	P. megasperma var. sojae
P. nicotianae var. parasitica	P. gonapodyides	P. mycoparasitica	P. melonis
P. palmivora	P. heveae	P. porri	P. mexicana
	P. lateralis	P. primulae	P. oryzae
	P. macrospora	P. spinosa var. lobata	P. phaseoli
	P. syringae	P. spinosa var. spinosa	P. quininea
	P. vesicula	P. verrucosa	P. richardiae
			P. vignae

Hosts

The following is not intended to be a complete host list for
each *Phytophthora* species. It is hoped however, that sufficient
hosts have been included for each species to give an indication
of the range of plants attacked by individual species of the
fungus.

Note: Only naturally infected hosts are included in this list.

P. *arecae*

  Areca catechu

P. *bahamensis*

  Rhizophora mangle

P. *boehmeriae*

  Boehmeria nivea, Conifers

P. *botryosa*

  Arerrhoa carambola, Hevea spp., *Vigna sesquipedalis*

P. *cactorum*

  Parasitizes over 83 genera (44 families), including: *Arbutus,
Conifers, Cornus, Cotonaster, Humulus, Juglans regia, Lilium, Lycopersicon, Malus, Panax, Phlox, Persea, Prunus, Pyrus, Rhododendron, Tulipa, Vicia,* etc.

P. *cambivora*

  Deciduous fruit trees and nuts (e.g. *Malus, Prunus, Pyrus*),
Castanea spp., Beech, Maple, etc.

P. *capsici*

  Capsicum annum (pepper), *Citrullus vulgaris* (watermelon),
Cucumis melo (honeydew melon), *Cucurbita maxima* (squash), *Cucurbita pepo* (pumpkin), *Cucurbita sativa* (cucumber), *Lycopersicon
esculentum* (tomato), *Macadamia integrifolium* (macadamia), *Phaseolus lunatus, Solanum melongena* (eggplant).

P. carica

   Ficus carica (fig)

P. castaneae

   Castanea crenata

P. cinnamomi

   Parasitizes over 900 different plants including: *Ananas comosa* (pineapple), *Camellia, Carica papaya* (pawpaw), *Cinnamomum burmanii, Eucalyptus* spp., *Grevillea, Hibiscus, Macadamia, Persea* (avocado), *Rhododendron,* etc.

For complete host list see:

ZENTMYER, G.A. 1979. Monograph of *Phytophthora cinnamomi*. Amer. Phytopathol. Soc., St. Paul., Minnesota, U.S.A. (In press).

ZENTMYER, G.A. & W.A. Thorn. 1967. California Avocado Yearbook, 51: 177–186.

P. citricola

   Antirrhinum, Ceanothus, Citrus spp., *Hibiscus* spp., *Humulus lupus* (hops), *Malus, Persea, Pieris japonica, Pinus, Rhododendron, Rubus, Schinus molle, Syringa,* etc.

P. citrophthora

   Citrus and related genera, *Morrenia odorata* (milkweed vine), *Pieris japonica, Rhododendron.*

P. colocasiae

   Colocasia esculenta (Dasheen), *Xanthosoma* spp.

P. cryptogea

   Callistephus, Conifers, Lycopersicon esculentum (tomato), *Petunia, Pseudosuga taxifolia* (Douglas fir), *Rhododendron, Tagetes, Zinnea,* etc.

P. cyperi

   Cyperus malaccensis

P. cyperi-bulbosi

   Cyperus bulbosus

P. drechsleri

   Albizia stipulata (Leguminosae), *Beta, Brassica rapa* (turnip), *Capsicum annum* (pepper), *Carica papaya* (pawpaw), *Carthamus tinctorius* (safflower), *Celosia plumosa* (Amarantaceae), *Chrysanthemum cinenaniifolium, Citrullus vulgaris* (watermelon), *Cucurbita melo* (honeydew melon), *Cucurbita pepo* (pumpkin), *Daucus carrota* (carrot), *Guayule* spp., *Gerbera* spp., (Compositae), *Lycopersicon esculentum* (tomato), *Malus* (apple), *Mucuna deeringiana* (Leguminosae), *Parthenium argentatum* (Compositae), *Pastinaca sativa* (parsnip), *Pelargonium zonale* (Geraniaceae), *Prunus* spp. (cherry, plum), *Solanum melongena* (eggplant), *Solanum tuberosum* (potato), *Schinus molle* (Anacardiaceae), *Senecio cruentus* (Compositae).

Useful references:

COTHER, E.J. 1975. Australian J. Botany, 23: 87-94.

P. erythroseptica

   Rubus, Solanum, Tulipa, Zantedeschia (calla-lily).

P. fragariae

   Dryas, Fragaria, Geum, Potentilla, Rubus.

P. gonapodyides

   Alnus, Rhododendron, Pyrus, etc.

P. heveae

   Brazil nut, Hevea (rubber), *Persea* (avocado), *Theobroma* (cacao).

P. hibernalis

   Citrus spp.

P. infestans

   Solanaceae, Lycopersicon esculentum (tomato).

P. inflata

   Ulmus sp.

P. lateralis

   Chaemaecyparis, Conifers, Rhododendron.

P. lepironiae

   Lepironia mucronata.

P. macrospora

   Avena, Hordeum, Oryza, Triticum, Zea, wild grasses.

P. meadii

   Hevea brasiliensis

P. megasperma var. *megasperma*

   Parasitizes 14 genera (9 families) including: *Althea, Brassica, Cheiranthus, Daucus,* deciduous fruit trees and nuts, *Medicago, Rhododendron, Spinacea,* etc.

P. megasperma var. *Sojae*

   Glycine max

P. melonis

   Cucumis sativa

P. mexicana

   Lycopersicon esculentum

P. mycoparasitica

   Rhizophora sp.

P. nicotianae var. *nicotianae*

Allium, Catharanthus, Citrus spp., *Lilium, Lupine, Lycopersicon, Nicotiana, Solanum, Vicia taba,* etc.

P. nicotianae var. *parasitica*

Parasitizes over 72 genera (42 families) including: *Ananas, Buxus* spp.,*Citrullus vulgaris, Colocasia, Cucumis melo, Hevea, Lycopersicon, Nicotiana, Petunia, Piper, Pyrus, Ricinus, Sesamum,* etc.

P. oryzae

Oryza seedlings.

P. palmivora

Parasitizes over 51 genera (29 families, approx. 138 different plants) including: *Allium* spp., *Carica papaya, Cocos nucifera, Citrus* spp., *Dieffenbachia, Ficus carica, Gossypium* sp., *Hevea* spp., *Hibiscus, Lycopersicon, Persea, Phaseolus, Theobroma,* etc.

For complete host list see:

CHEE, K.H. 1969. Hosts of *Phytophthora palmivora.* Rev. Appl. Mycol. 48: 337-44.

CHEE, K.H. 1974. Pages 81-87 in, *Phytophthora* disease of cocoa (P.H. Gregory, ed.). Longman Group. Ltd., London & N.Y.

P. phaseoli

Phaseolus lunatus.

P. porri

Allium spp.

P. primulae

Primula (polyanthus).

P. quininea

   Cinchona sp.

P. richardiae

   Zantedeschia (Calla-lily).

P. syringae

   Syringa, Pyrus.

P. spinosa var. *lobata*

   Rhizophora sp.

P. spinosa var. *spinosa*

   Australian pine, buttonwood, *Cocos* spp., *Rhizophora mangle,* white mangrove.

P. verrucosa

   Lycopersicon, Meconopsis sp.

P. vesicula

   Australian pine, buttonwood, *Cocos* spp., *Distichilis, Prunus laurocerasus,* sea grape, *Thalassia,* white, black, and red mangrove, etc.

P. vignae

   Vigna unquiculata (cowpea).

Diseases caused by *Phytophthora* spp.

P. arecae

   Foliage blight of *Areca catechu; Areca* palm pod blight.

P. bahamensis

   Leaf litter decay in mangrove swamps.

P. boehmeriae

Leaf spots on *Boehmeria nivea*.

P. botryosa

Hevea black stripe, *Cacao* fruit rot.

P. cactorum

Seedling damping-off; fruit rots; root rots of a wide variety of plants (see host list for names of plants); collar rot of apples; leather rot of strawberries; crown rot of walnut; stem rot and wilt of snapdragon.

P. cambivora

Ink disease of chestnut, crown and root rot of fruit trees; root rot of forest trees.

P. capsici

Stem and fruit rot of peppers; buckeye of tomato fruit; soft rot of muskmelon and watermelon.

P. carica

White rot of fig fruits.

P. castaneae

Dieback of chestnut.

P. cinnamomi

Bark stripe canker of cinnamon; root rot of avocado; heart rot of pineapple; Jarrah dieback of *Eucalyptus* spp. etc.

P. citricola

Fruit rot of citrus; shoot blight and stem dieback of *Pieris japonica*; root rot and dieback of *Rhododendron*.

P. citrophthora

Brown rot of citrus; foot rot or gummosis of citrus; shoot blight and stem dieback of *Pieris japonica*; stem dieback of *Rhododendron*.

P. colocasiae

Leaf blight of *Colocasia.*

P. cryptogea

Root rot of forest trees; root rot of ornamentals (see host list for types of plants attacked).

P. cyperi

Red disease or tortoise shell disease of Shichito Mat grass.

P. cyperi-bulbosi

Leaf spot of *Cyperus* sp.

P. drechsleri

Tuber rot of potato; buckeye of tomato; soft rot of muskmelon and watermelon; root rots of forest trees; root rots of ornamentals; crown rot of cherry trees.

P. epistomium

Leaf litter decay in mangrove swamps.

P. erythroseptica

Pink rot of potato tubers; root rots of ornamentals.

P. fragariae

Red stele of strawberry; root rot of ornamentals.

P. gonapodyides

Fruit rot of apples; twig blight of alder (*Alnus* sp.); root rot of *Rhododendron.*

P. heveae

Pink pod rot and black stripe of *Hevea;* root rot of avocado.

P. hibernalis

Spots on fruit, leaves, and twigs of *Citrus* spp.

P. infestans

Late blight of potato and tomato.

P. inflata

Root rot of elms.

P. lateralis

Root rot of *Chaemaecyparis* (Port-Orford cedar); root rot of *Rhododendron*.

P. lepironiae

Stem blight of *Lepironia* sp.

P. macrospora

Stem blight of cereals and grasses.

P. meadii

On leaves, fruits, and bark of *Hevea*.

P. megasperma var. *megasperma*

Destructive decay of stem and roots of numerous forage, vegetable and ornamental plants (see Host List for names of plants attacked); crown and root rot of deciduous fruit trees.

P. megasperma var. *sojae*

Root and stem rot of soybeans.

P. melonis

Cucumber fruit rot.

P. mexicana

Tomato fruit rot.

P. mycoparasitica

Leaf litter decay in mangrove swamps.

P. nicotianae var. *nicotianae*

Buckeye rot of tomato fruit; fruit rot of citrus; stem rot of lupine.

P. nicotianae var. *parasitica*

Black shank of tobacco; buckeye of tomato; soft rot of muskmelon and watermelon; root rot of a wide variety of plants (see Host List for names of plants attacked).

P. oryzae

Blight of rice seedlings.

P. palmivora

Fruit rot and patch canker of *Cacao*; coconut bud rot; fruit rot, shoot rot, leaf fall, and patch canker of *Hevea* (rubber); fruit and stem rot of papaya; rot of orchids; fruit rot of cotton; root rot of pineapple; dieback of hibiscus; damping-off of seedlings.

P. phaseoli

Downy mildew of lima beans.

P. porri

Leaf and stem blight of onions; white tip of leeks.

P. primulae

Stele discoloration and root rot of *Primula*.

P. quininea

Destructive root and collar rot of *Cinchona* spp.

P. richardiae

Root rot and decay of Calla-lily.

P. syringae

Lilac blight.

P. spinosa var. *lobata*

   Leaf litter decay in mangrove swamps.

P. spinosa var. *spinosa*

   Leaf litter decay in mangrove swamps.

P. verrucosa

   Root rot of tomato and *Meconopsis* spp.

P. vesicula

   Leaf litter decay in mangrove swamps.

P. vignae

   Stem and root rot of cowpea.

Section 2:

Isolation Techniques

2A: Non-specific hosts for baiting

Phytophthora spp.

A. Preparation of diseased tissues for *Phytophthora* isolations

Successful isolation of *Phytophthora* depends on the careful selection and handling of diseased tissues.

I. Isolation from above-ground plant parts

1. Select stems and twigs with actively progressing lesions,
   or diseased tissue that has not been desiccated or corked-
   off from healthy tissue.

2. Diseased tissues close to the soil should be washed and
   the outer layers of the bark stripped off to remove any
   saprophytic *Pythium* spp. that may be present.

3. Surface sterilization is not necessary since most selective antibiotic media will inhibit bacteria and common
   saprophytic fungi e.g. *Alternaria, Fusarium, Trichoderma,*
   etc.

4. Inverting the agar over small pieces of diseased tissue
   facilitates isolation of pure cultures (see page 281).

II. Isolations from roots and soil

The prevalence of *Pythium* make isolations of *Phytophthora* from
roots and soil difficult. The following methods have been found
to offer varying degrees of success against *Pythium* spp.

1. Hymexazol medium for inhibition of *Mortierella* and *Pythium.*

2. Multiple isolations.

3. Use of selective hosts.

4. Establishment of cultures from zoospores.

1. Hymexazol medium for inhibition of *Mortierella* and *Pythium*

Media have been developed that selectively inhibit *Pythium* and

Mortierella, and have no effect on *Phytophthora.* (See pages 61-62 for details of media). These media thus allow direct isolation of *Phytophthora* from soil without the fear of *Pythium* spp. overgrowing the *Phytophthora* cultures.

2. Multiple isolations

Essentially this technique consists of placing quantities of soil or debris on selective media, incubating for several days, washing off the soil or debris, and selecting the easily recognizable *Phytophthora* colonies from the *Pythium* colonies. See pages 53-63 for selective media suitable for this technique.

3. Use of selective hosts

Hosts such as apples, lemons, avocados, strawberries, hibiscus, lupins, etc. can be used to bait *Phytophthora* in infested soil. See details of these techniques described on pages 26-49.

4. Zoospores technique

a) Wash diseased roots thoroughly by placing them in a 5 inch sieve in a 600 ml beaker of water containing 1 ml of Tween 20. Run cold water over the roots until all traces of foaming from the wetting agent have disappeared.

b) Incubate the roots in petri plates containing 15 ml of 1/3 pond water, 2/3 sterile water, or in Schmitthenner's salt solution for Sporangia formation (page 107).

c) When Sporangia form, induce Zoospore differentiation by placing the petri plate in a refrigerator for 30 minutes. Bring back to room temperature.

d) Collect the Zoospores with a micropipette and place on a selective antibiotic agar medium. Invert the agar to spread the Zoospores.

e) Transfer colonies developing from the Zoospores

within 40-48 hrs. before any *Pythium* that may be present
overrun the plates. Alternatively, use the Hymexazol selective medium (pages 61-62).

SCHMITTHENNER, A.F. 1973. Proc. First Woody Ornamental Disease Workshop, 1:
94-110.

Isolation Techniques

Useful References:

1. SCHMITTHENNER, A.F. 1973. Isolation and identification methods for *Phy-*
   tophthora and *Pythium*. Proc. First Woody Ornamental Disease Workshop.
   Univ. of Missouri, Columbia, Missouri, pages 94-110.

2. SCHMITTHENNER, A.F., and J.W. HILTY. 1962. A modified dilution technique
   for obtaining single-spore isolates of fungi from contaminated material.
   Phytopathology, 52: 582-583.

3. TSAO, P.H. 1970. Selective media for isolation of pathogenic fungi. *In,*
   Annual Review of Phytopathology (J.G. Horsfall & K.F. Baker eds.). Ann.
   Reviews Inc., Palo Alto, Calif. 8: 157-186.

4. WATERHOUSE, G.M., and D.J. STAMPS. 1969. *Phytophthora* and *Pythium,* pages
   92-102. *In,* Isolation methods for microbiologists. Soc. Appl. Bacteriol.
   Tech. Ser. No. 3 Academic Press, London, England.

B. Criteria for the ideal bait in a *Phytophthora* bioassay are:

   1. Susceptibility to most if not all root-infecting *Phytoph-*
      thora spp.

   2. High sensitivity, especially when inoculum levels are low.

   3. The bait should be of reasonable size.

   4. The bait should be fairly inexpensive.

   5. Ready availability of baits, geographically and seasonally.

   6. Convenience of use in setting up an assay and in subse-
      quent isolating procedures.

DANCE, M.H., F.J. NEWHOOK, and J.S. COLE. 1975. Plant Dis. Reptr. 59: 523-527.

C. Baits for *Phytophthora*

The following pages contain descriptions of a variety of baits useful for *Phytophthora* bioassays. Although only a few of these meet the above criteria, it should be noted that several of them, when used in conjunction with each other, considerably extend the range of *Phytophthora* spp. that can be detected in field soils.

Note:

Some *Phytophthoras*, e.g. *P. palmivora* are difficult to detect when the soil is dry. Dormancy of the inoculum can be broken by wetting soil samples for at least 5 days before testing (Okaisabor, Nigerian Agr. J. 6: 85-89, 1969).

Baiting Methods

I. Apple

1. Pack soil into 1/2 inch hole bored into an apple with a cork borer, so that the soil almost completely fills the hole.

2. Fill the remaining space in the hole with distilled water.

3. Seal the hole with tape.

4. Incubate at 18-26°C.

5. *P. cinnamomi*, if present will cause a firm dry rot of the apple tissue in 5-10 days.

CAMPBELL, W.A. 1949. Plant Dis. Reptr. 33: 134-135.

Notes:

a) This method is not successful for *P. megasperma* isolations.

b) Lupin radicles are more sensitive than the apple technique
    for the detection of *P. cinnamomi*.

II. Cacao pods

1. Remove an 8 mm plug from an unripe pod of cacao (*Theo-
    broma cacao*).

2. Push into the hole a wedge of diseased bark or wood
    about 10 mm wide and 20 mm long, until flush with the
    surface. Up to six wedges can be inserted in a single
    pod.

3. Seal the pod in a polyethylene bag and incubate at room
    temperature $(26-30^{\circ}C)$.

4. In 4-5 days, a brown discoloration develops around each
    plug.

5. An almost pure culture of *Phytophthora* can be obtained
    by taking a small piece of tissue from under the surface
    of the pod and plating it on soil-extract agar. Cultures
    are purified by subculturing on pea extract agar.

6. This method can also be used to recover *P. nicotianae*
    var. *parasitica* as well as *P. palmivora* and *P. meadii*,
    by packing soil from diseased areas into holes (20 mm
    diam.) made in cacao pods.

7. This technique has the advantage of being conveniently
    employed in the field.

Note:

The rot should be firm; a soft rot indicates colonization of
the pod by saprophytic organisms.

CHEE, K.H. and K.M. FOONG. 1968. Plant Dis. Reptr. 52: 5.

II(a). Freshly germinated cacao seedlings (1-2 cm radicles) can
also be used to detect the presence of *P. palmivora* in infested
soil.

BOCCAS, B. 1975. O.R.S.T.O.M. Brazzaville, Congo.

III. Citrus leaf bait technique modified for irrigation water

1. "Feather" whole leaves of Rough lemon (*Citrus jambhiri*)
   by removing 2 mm strips from the leaf margins and making
   several short cuts (10 mm apart) perpendicular to midrib.

2. Tie a string to the petioles of three "feathered" leaves
   so that individual leaves are approximately 15 cm apart.

3. Float three-leaf sets on irrigation water at field sites
   for at least 5 hrs.

4. Collect leaves in sterile bottles, along with 50 ml of
   irrigation water to keep leaves moist during transport
   to the laboratory.

5. Float leaves for an additional 36 hrs. (24°C) on sterile
   distilled water before attempting identification of colonies.

THOMSON, S.V. and R.M. ALLEN. 1974. Plant Dis. Reptr. 58: 945-949.

IV. Cucumber Technique

1. Place 40-60 g of soil in a sterilized wide-mouthed glass
   jar.

2. Add tap water until level is approx. 1 cm above saturated soil.

3. Place a fresh cucumber fruit, previously surface sterilized with ethyl alcohol, in an upright position in the
   jar.

4. Less than half the cucumber should be in contact with
   the soil-water mixture in the jar.

5. Incubate at 20°C for 4-7 days.

6. As soon as signs of infection are observed, remove fruit, surface-sterilize with ethyl alcohol, place in sterilized empty glass jars and incubate at 20°C.

7. Rather pure cultures of either *Phytophthora* or *Pythium* are obtained by placing small pieces from the infected fruit on acidified corn meal agar.

Note:

Young cantaloupe, muskmelon and watermelon fruits can be used, but squash is not comparable.

BANIHASHEMI, Z. 1970. Plant Dis. Reptr. 54: 261-262.

V. Hemp or safflower seeds method

1. Collect moist soil samples and place in plastic bags in a cold room (4°C), until ready for use.

2. Take small samples from each bag, grind and place in petri plates to a depth of 1-1.5 cm.

3. Flatten with a glass rod, or by hand.

4. Place a nylon screen on the soil surface so that it adheres completely.

5. Place 25 seeds of either hemp (*Cannabis sativa* L.) or safflower (*Carthamus tinctorius* L.), previously boiled for 5 minutes on the nylon screen. Do *not* overboil the seeds.

6. Add sterile water to saturate the soil and to cover the seeds with a layer of water.

7. Incubate at 20°C.

8. Remove seeds 2-3 days later with sterile forceps and transfer to lima bean extract agar, or sterile water.

9. Incubate at 20-25°C.

10. *Pythium* and *Phytophthora* grow out rapidly from the seeds.

Notes:

a) Soybean, cowpea, and rape seeds are unsuitable.

b) There is less chance of contamination by other soil fungi
    if seeds are placed on the nylon screen.

c) This method does not require chemicals or special equipment.

FATEMI, J. 1974. Phytopath. Mediterr. 13: 120-121.

VI. Lemon-orange technique

1. Place an orange or lemon in a can provided with a drain
    hole and wire bail.

2. Bury can 6-12 in. under the soil surface, with the wire
    bail protruding above the soil for easy removal of can.

3. Depending on soil temperature, 4-10 days after irriga-
    tion or rain, the fruit can be pulled out and isolation
    made from fruit showing brown rot.

4. *P. citrophthora*, *P. parasitica*, *P. syringae*, and *P. hi-
    bernalis* can be isolated by this technique.

KLOTZ, L.J. and T.A. DeWOLFE. 1958. Pl. Dis. Reptr. 42: 675-676.

VII(a) Lupin technique

1. Surface sterilize lupin seeds (*Lupinus augustifolius* L.),
    by immersing in 50% alcohol and then soak overnight in
    distilled water.

2. Plant the seeds in vermiculite or perlite which has been
    pre-moistened with 1 ml of water for each 2 cc of vermi-
    culite.

3. Place trays in the greenhouse at 22-26°C, for 48-60 hrs.

4. Select seedlings with uniform radicles about 2 in. (48 mm) long.

5. Remove adhering particles of vermiculite by washing in a muslin-covered beaker under running water.

6. Place 3/4" (20 mm) layer of soil to be tested in a plastic container and add tap water to the top of the container.

7. Place the radicles so that the radicle tips come into contact with the surface of the soil.

8. If the soil sample is dry, soak the soil for 28 hrs before using.

9. Leave the radicles in the container for 48 hrs at room temperature.

10. Brown or orange-colored lesions develop behind the root tip.

11. Plate radicles on an antibiotic agar medium to isolate and identify *Phytophthora* spp. present.

12. This technique has been used successfully to bait *P. boehmeriae, P. cinnamomi, P. megasperma, P. nicotianae* var. *parasitica,* and *P. syringae.* It is not sensitive to *P. cactorum* and *P. cryptogea.*

Advantages

a) Seedlings are ready for soil testing 60 hrs. after sowing.

b) Seeds usually give close to 100% germination.

c) Lesions develop within 24 hrs even with low inoculum densities.

CHEE, K.H. and F.J. NEWHOOK. 1965. New Zealand J. Agr. Res. 8: 88-95.

Note:

This technique has also been used successfully to bait *P. cambivora, P. citricola, P. hibernalis, P. nicotianae* var. *nicotianae,* and *P. richardiae*. (Dance *et al*. Plant Dis. Reptr. 59: 523-527, 1975).

Lupin radicles grow rapidly in the soil solution and can reach the bottom of the container in 48 hrs. If the fungus is not detected before this time, the experiment has to be repeated. Also, identification of the fungus is slow since the radicles have to be plated on antibiotic media resulting in the frequent masking of *Phytophthora* by rapidly growing *Pythium* spp.

The following method overcomes these difficulties for baiting *P. cinnamomi*.

VII(b) Lupin technique

1. Soak blue lupin (*Lupinus augustifolius* L.) seeds in tap water for 24 hrs and allow to germinate in wet vermiculite.

2. When radicles are between 1 and 2 cm long, cut off with a sharp scalpel and float the excised radicles in 2-3 cm of Cu^{2+} free water covering 30 ml soil in 90 mm petri dishes.

3. Examine the excised radicles under the microscope 2-3 days after immersion in soil leachate.

4. Sporangia of *P. cinnamomi* can usually be observed on roots floating in infested soil. Some *Pythium* can also be observed.

5. The disadvantage of this method is that soil bacteria and *Fusarium* spp. occasionally infect radicles, suppressing sporangia of *P. cinnamomi*.

MARKS, G.C., F.Y. KASSABY, and S.T. REYNOLDS. 1972. Austral. J. Botany 20: 141-154.

VIII. Pear Technique

1. Place green pears in a slurry of 50 ml of test soil and

100 ml distilled water.

2. When firm brown lesions appear on the fruits (6-7 days),
   place pieces of the infected tissue on Difco cornmeal
   agar.

3. *P. cactorum*, *P. cambivora*, *P. cryptogea*, and *P. drechsleri* have been isolated by this technique.

Notes:

a) This technique is more convenient than the apple technique
    (page 27), since holes are not required to be bored in the
    fruit. Also, *Pythium* spp. can outgrow and mask *Phytophthora*
    in apples.

b) Suspending fruit tree seedlings in soil slurries were found
    to be unsatisfactory due to the frequent invasion of the
    rootlets by other fungi.

McINTOSH, D.L. 1964. Can. J. Botany 42: 1411-1415.

IX. Pineapple technique

1. Add infested soil or diseased plant tissue to tap water
   contained in paraffin-paper cups.

2. Partially immerse previously rooted pineapple crowns into the paper cups.

3. Alternately, young heart leaves of pineapple crowns or
   slips are strung over the top of the paper cups, so that
   the white basal tissue of each leaf is immersed in the
   water.

4. Incubate at 18-27°C.

5. Resulting infections are nearly free of secondary in-
   vaders.

6. Transfer of the infected bait tissue to a Pimaricin medium increases recovery of Phythiaceous fungi.

KLEMMER, H.W. and R.K. NAKANO. 1962. Phytopathology 52: 955-956.

A similar isolation technique for *P. cinnamomi* using pineapple
heart leaves or rooted crowns is described by Anderson,   E.J.
1951. Phytopathology 41: 187-188.

X.   Pine needle technique

   1. Partially immerse or float basal segments or whole nee-
      dles of *Pinus radiata* or *Cedrus deodara* in water above
      soil samples contained in either open-topped beakers or
      waxed paper cups.

   2. Incubate at 16-18°C and 20-23°C.

   3. After 3 days, remove all needles regardless of whether
      symptoms can be observed or not.

   4. Surface sterilize with 1% sodium hypochlorite for 30
      secs, rinse three times with sterile distilled water,
      blot dry and plate on prune extract agar with 10 ppm
      pimaricin.

   5. Most *Phytophthora* spp. infect the needles only at the im-
      mature bases or at cut ends.

Notes:

a)   *Pythium* spp. also invade needle bases and cut ends, making
     Phytophthora detection difficult.

b)   Incubation temperatures influence the recovery of *Phytophtho-*
     ra spp. e.g. *P. citricola* can be recovered at 20-23°C, where-
     as only *P. hibernalis* is recovered at 16-18°C.

c)   In dry summer soils, the sensitivity of the assay can be in-
     creased by substituting autoclaved or millipore-filtered
     peptone/soil extract (1 g Bacto-peptone incubated 1-3 days
     in 100 g soil, extracted with 1 liter distilled water), in
     place of distilled water. Alternatively, a 1:2 mixture of
     pond and distilled water can replace the peptone-soil ex-
     tract medium.

d) This technique has approximately 40% greater efficiency than the lupin technique (page 30). When the pine needle technique is used in conjunction with the lupin technique, most *Phytophthora* spp. can be detected.

e) Isolates of the following *Phytophthora* spp. have been detected with the pine needle technique: *P. boehmeriae, P. cactorum, P. cinnamomi, P. citricola, P. citrophthora, P. cryptogea, P. drechsleri, P. heveae, P. hibernalis, P. megasperma* var. *megasperma, P. megasperma* var. *sojae, P. nicotianae* var. *nicotianae, P. nicotianae* var. *parasitica, P. palmivora, P. richardiae,* and *P. syringae.*

Note:

The lupin technique (VIIa and VIIb, pages 30 and 32) does not detect *P. cactorum* in field soils; the above techniques does detect this species.

DANCE, M.H., F.J. NEWHOOK, and J.S. COLE. 1975. Plant Dis. Reptr. 59: 523-527.

XI. Strawberry technique

1. Take a sample of field soil (0-5 cm deep).

2. Mix soil with water (2 soil : 1 water) in a beaker.

3. Let the soil sediment, and then pour supernatant in petri plates.

4. Place young green strawberries, previously treated with Benomyl (1 g Benomyl/liter of water), in the supernatant.

5. If *Phytophthora* is present, strawberries will be covered with white mycelium in 7 days.

6. *P. cactorum, P. capsici, P. cinnamomi, P. citrophthora, P. fragariae,* and *P. nicotianae* var. *parasitica,* can be successfully baited by this method. *P. cryptogea* can be isolated at lower temperatures (18°C).

7. *P. infestans* is not isolated by this technique.

Note:

This is not a quantitative method.

MOLOT, P.M. and J.G. NOURRISEAU. 1974. Annales Phytopathol. 6: 308-309.

MOLOT, P.M. and A. BEYRIES. 1976. Annales Phytopathol. 8: 91-94.

Section 2:

Isolation Techniques

2B: Hosts for baiting specific *Phytophthora* sp.

I. Hosts for isolation of *P. cinnamomi* from soil

A. Fruit trap

 1. Half embed firm *Fuerte* avocado fruit in soil previously
    placed in containers.

 2. Flood soil with water.

 3. In 3-6 days, infections appear as brown, firm, circular
    spots at the water line.

 4. Optimum period for leaving fruit in the mud trap is 48-
    96 hrs.

 5. Optimum temperature for fruit infection is 27°C.

 6. Other fungi rarely invade firm unwounded fruit.

B. *Persea indica* seedlings

 1. Plant seedlings of *Persea indica* (a small fruited relative
    of the avocado), in infested soil.

 2. Seedlings develop stripe cankers on the stem in 2-3 days.

 3. This is a sensitive method for detecting the presence of
    P. cinnamomi in soil.

C. Avocado tree roots

 1. Dip small, washed, infected roots, from diseased avocado
    trees in 70% alcohol for 15 secs and blot dry.

 2. Then place roots on cornmeal agar plates.

 3. In 48 hrs. at 20-24°C, the fungus can easily be identified.

 4. Later other fungi overgrow *P. cinnamomi* colonies.

ZENTMYER, G.A., J.D. GILPATRICK, and W.A. THORN. 1960. Phytopathology 50:
87.

Notes:

a) "Cuke" avocados (improperly fertilized avocados), can also
   be effectively used as baits.

b) An evaluation of the various methods for the isolation of *P.
   cinnamomi* indicated that all the above methods were equally
   effective in detecting *P. cinnamomi* in infested soil. (Brod-
   rick, H.T., G.A. Zentmyer, and R. Wood. 1976. Avocado Yearbook 1975-76: 87-91.

II. *P. fragariae:* Strawberry runners technique

   1. Grow mother plants of *Fragaria vesca* clone VS, in the
      greenhouse at approx. 15°C in a peat/sand compost in
      125 mm plastic pots.

   2. Remove runners at 1-2 expanded leaf stage and root in
      trays of fresh compost in a mist propagating chamber.

   3. Use runners 3-8 weeks later, depending on size.

   4. Wash roots free of compost before using.

   5. Alternatively bait plants may be grown from seed.

   6. Place in infested soil at different dilutions.

   7. After 5 weeks, wash the root systems of the bait plants
      free of soil.

   8. Examine for symptoms of red stele and the presence of
      oospores.

   9. The number of plants infected at each dilution level can
      be used to prepare a 'most probable number' estimate of
      the number of 'infective units' of the fungus in the
      original soil samples.

Note:

Non-infection of bait plants does not necessarily indicate the
absence of the pathogen in the soil.

DUNCAN, J.M. 1976. Trans. Brit. Mycol. Soc. 66: 85-89.

III. Isolation of *P. lateralis* from streams

A. Styrofoam boat technique

1. Use *Chaemaecyparis* (cedar) seedlings between 3-10 in (8-25 cm) tall.

2. Carefully shake the seedling loose from planting soil.

3. Place the seedlings upright in a small styrofoam boat so that the roots extend below the boat bottom into the water. Make the boat of 1 1/2 in styrofoam (6 in long and 4 in wide, 15 x 10 cm), and split it vertically down the mid-line. Clamp the seedling in place in the center of the boat by joining the two halves of the boat with rubber bands.

4. Moor the boat at the test site by attaching a wire to a stake flag (Fig. 1A).

5. Let the boat float in the stream for up to 10 days.

6. Remove the seedling after assaying for the required period of time, and replant in wet peat moss.

7. Observe seedling for symptoms of *Phytophthora* root rot.

8. This method is suitable for large streams in the forest habitat.

B. Planting cells technique

1. Grow cedar seedlings in small (3/4 x 4 1/2 in, 2 x 12 cm) planting cells.

2. Place the plant cells in trays of shallow water so that the roots of the young seedlings eventually grow through the perforated cell bottoms into the water. The seedlings are then ready for baiting *Phytophthora*.

3. Place the entire cell together with the seedling and exposed roots, in an upright position in a monitoring site.

Fig. 1: Exposure of roots of living seedlings for the assay of *Phytophthora lateralis*. A. Soil free seedling floated in small plastic boat, suitable for larger streams with fluctuating flow. B. Intact seedling in planting cell, suitable for very small streams and surface water. Reproduced by permission of the author.

Anchor by driving a flag wire into the planting medium
and out the bottom of the cell deep into the floor of
the stream (Fig. 1B).

4. If the cells are used to monitor the surface of wet
   soils, cover the roots of the seedlings lightly with wet
   peat moss to reduce drying.

5. After exposure, place cells in foil-darkened test tubes
   to keep the exposed root moist during incubation.

6. *Phytophthora* infested seedlings are killed.

7. This method allows a more detailed study for the presence
   of *P. lateralis*. It is suitable for small rivulets and
   for surface water or wet soils in the rainy season.

Note:

These techniques of monitoring streams, provide information of
the onset of zoospore production, and indirectly of the presence
of active *Phytophthora* within the watershed.

ROTH, L.F. 1977. Phytophthora Newsletter 5: 12-13.

IV(a)   Isolation of *P. megasperma* from soils

1. Surface-sterilize alfalfa seeds (var. *Vernal*) with 3% sodium hypochlorite for 30 secs.

2. Thoroughly wash the seeds in sterile water and germinate
   on moist filter paper in petri plates.

3. Germination occurs between 24-36 hours at 22-24°C, and
   the seedlings are ready for use after 72 hours, when
   radicles are 6-10 mm long.

4. Place about 30 ml of soil in each 90 mm petri dish and
   cover to a depth of 2-3 mm with 35-40 ml of distilled
   water.

5. Crush hypocotyl and radicle tissues at 2 or 3 points with
   a pair of fine tweezers, and float six of these injured

seedlings in water in each petri plate. Make sure that
seedlings are not totally submerged.

6. Crushing the hypocotyls and radicles slows the growth
   and maturation of the seedlings.

7. Incubate at 20-22°C for 6 days.

8. First evidence of infection appears within 2-4 days,
   when the cotyledons become water-soaked and sporangia of
   P. megasperma can be observed on the cotyledons and hypocotyls.

9. Infected seedlings usually become quite flaccid.

10. In sterile soils, or soils taken from areas not infested
    with *P. megasperma*, the seedlings remain turgid and fresh
    for 7 to 10 days.

MARKS, G.C. and J.E. MITCHELL. 1970. Phytopathology 60: 1687-1690 (see also,
Antibiotic media, pages 57-63).

IV(b)   Isolation of *P. megasperma* var. *sojae* from field soil

1. Fill 12 oz paraffin coated paper cups with field soil to
   be tested, and plant soybean seeds 2 cm below the soil
   surface.

2. Punch a 5 mm hole in the side of the cup 2 cm below the
   level of the seeds.

3. Fill the cup with distilled water.

4. Incubate the cups in the greenhouse at 20-30°C for 7 days
   and note seedling emergence.

5. Plate rotted seedlings on a selective antibiotic medium
   (Pages 57-63), for isolation of the pathogen.

6. Seeds of soybean varieties resistant to different races
   of the pathogen can be planted in marked off sections in
   the same cup. This will give an indication of the pathogen's race, e.g. plant Harosoy and Harosoy 63 in each
   half of the cup; if all seedlings emerge healthy, then
   no pathogenic *Phytophthora* is present; if only Harosoy
   63 remain healthy, then race 1 of the pathogen is prob-

ably present; if none of the seedlings emerge healthy, then use other soybean varieties to determine rare specificity.

PAXTON, J.D. 1976. Proc. Amer. Phytopathol. Soc. 3: 239.

V. Isolation of *P. nicotianae* var. *parasitica* (*P. parasitica*)

A. Carnation petals technique

1. Remove young, immature and etiolated petals from carnation buds before pigmentation occurs, and keep fresh on moist filter paper.

2. Place 0.02 ml of soil suspension on the inner side of the petals and incubate at 24°C.

3. Two days later, infected petals become water-soaked and soft, and abundant sporulation occurs along the edges of the petals.

4. Healthy baits remain fresh for about 7 days.

5. Eventually, other soil fungi will infect the petals if left for a long period of time.

PONCHET, *et al.* 1972. Annal. Phytopathol. 4: 97-108.

B. Citrus leaf pieces technique

1. Place soil samples of 100 ml volume in waxed cardboard containers.

2. Add sufficient rain water to the soil samples to flood the soil to a depth of 1-2 cm.

3. Cut 3-5 mm square pieces from mature citrus leaves and float on the surface of the water.

4. Allow to remain at room temp. (22-28°C) for 3-4 days.

5. Abundant sporangia develop along the cut edges of the leaf pieces.

6. Subculture of *P. parasitica* from infected leaf pieces is
   easily accomplished by submerging infected leaf pieces
   in molten cornmeal agar containing 10 μg/ml pimaricin,
   200 μg/ml Vancomycin and 100 μg/ml PCNB (pentachloronitrobenzene).

Note:

Sporangia are not formed on edges of leaf pieces that sink, or
along the natural margins of whole leaves.

This method is more convenient than the fruit-trapping technique to be indexed.

GRIMM, G.R. and ANN ALEXANDER. 1973. Phytopathology 63: 540-541.

C. Hibiscus seedling technique

1. Place young *Hibiscus* seedlings in soil.

2. *P. parasitica* and *P. palmivora* will cause collar cankers
   in 3 to 4 days.

3. *P. cinnamomi* will not infect seedlings.

FOLLIN, J.C. and B. BOCCAS. O.R.S.T.O.M. Brazzaville, Congo.

VI. Isolation of *P. palmivora* from soil

A. Cocoa pods

1. Surface sterilize whole pods with 1% sodium hypochlorite.

2. Lay singly in petri dishes with 20 g of soil kept saturated with distilled water.

3. Incubate without lids at 21-32°C, and check daily.

4. When rots are observed, incubate pods overnight at high
   humidities to induce sporulation.

B. Soil in Cocoa pods

1. Bore holes 50 mm from the proximal and distal ends of
   mature but unripe cocoa pods. Drill holes with a 12 mm
   cork-borer sufficiently deep so as to reach the mesocarp
   or endocarp layers.

2. Place 3 g of well-mixed soil in each hole.

3. Close the inoculated areas by replacing part of the cyl-
   inder of the pod tissue and seal with petroleum jelly,
   or alternatively, cover with facial tissue and seal the
   edges of the tissue with petroleum jelly.

4. Add water 3 times each day (for 2 days), through the
   facial tissue until the soil is fully saturated.

Notes:

a) Sealing the holes with pod tissue and vaseline gives better
    results than covering holes with facial tissue and irrigat-
    ing periodically.

b) The holes must be drilled to the endocarp layer for best re-
    sults. Shallow soil holes result in less efficient detection
    of *P. palmivora*.

C. Cocoa pod plugs in petri dishes

1. Cut plugs of cocoa tissue from pods using a 12 mm cork-
   borer.

2. Cut each plug sufficiently deep (10-12 mm) so that it
   reaches down to the endocarp.

3. Place ten plugs, epidermis uppermost, in petri dishes together with 20 g soil.

4. Add about 30 cm^3 sterile distilled water to the dishes
   until the plugs are half-submerged. Add additional water
   as necessary to maintain this level.

48 ISOLATION TECHNIQUES

Notes:

The advantages of this method are:

a) Only a few pods are required to bait large number of soil samples.

b) Profuse sporulation usually occurs just below the top of the plug, and can easily be distinguished from the spores of other fungi.

c) Less chance of contaminants when compared to the whole pod method.

D. Cocoa pod plugs in vials

1. Place soil samples in plastic vials (50 mm diameter, 90 mm high).

2. Place 5 cocoa plugs, epidermis uppermost, in each vial and add 10 cm^3 of sterile distilled water.

NEWHOOK, F.J. and G.V.H. JACKSON. 1977. Trans. Brit. Mycol. Soc. 69: 31-38.

Notes:

a) The plug method (C above) has been reported to be the most efficient and rapid method for detection of *P. palmivora* (Newhook & Jackson, 1977). This method can detect inoculum in soil diluted 1:1000, compared to 1:100 sensitivity for soil placed in holes drilled into pods.

b) Conflicting results are reported for the baiting of *P. palmivora* with whole undamaged cocoa pods. Some investigators claim success with this method, while others have failed to detect *P. palmivora* in infested soil. Newhook and Jackson (1977) suggest that this may be a reflection of the differences between varieties of cocoa pods used and the level of soil-borne inoculum in a particular location.

c) Although reports indicate that isolations are best when conducted with soil close to the trunk of the tree, Newhook and Jackson (1977) have found that *P. palmivora* can be detected more or less uniformly through plantation soil.

Section 2

Isolation Techniques

2C: Antibiotic media for the general isolation

of *Phytophthora* spp.

A. Properties of antibiotics commonly used in *Phytophthora*
   media

   Benlate - (See Benomyl)

   Benomyl - (Benlate, Tersan 1991); methyl 1 - (butyl carba-
           moyl) - 2 - benzimidazol carbamate. A systemic
           fungicide good against most non-pythiaceous fungi.

   Chloramphenicol - a broad spectrum antibiotic effective
           against gram positive and gram negative bacteria,
           rickettsiae and some viruses.

   PCNB (Pentachloronitrobenzene) - (Brassicol, Terraclor) -
           primarily a soil fungicide.

   Penicillin "G" - effective against gram positive and gram
           negative cocci, gram positive bacilli, spirochae-
           tes, and actinomycetes. Not effective against gram
           negative bacilli.

   Pimaricin (Myprozine) - effective against most non-pythia-
           ceous fungi.

   Polymixin B sulfate - effective against gram negative bac-
           teria.

   Rose Bengal - effective against most bacteria; its toxicity
           is increased when exposed to light.

   Streptomycin - effective against many gram negative and some
           gram positive bacteria.

   Vancomycin (Vancocin) - effective against most gram positive
           bacteria, particularly staphylococci resistant to
           most other antibiotics.

Note:

Sensitivity to antibiotics varies between different isolates of *Phytophthora*. See: Ersek, T. (1975). The sensitivity of *Phytophthora infestans* to several antibiotics. Pflanzenkrankheiten/ Pflanzenschutz Z. 82: 615–617.

B. Possible sources of antibiotics for selective media

   Anti-bacterial, not strongly inhibitory to *Phytophthora*

      Vancomycin Hydrochloride (one trade name is Vancocin) (not generally available commercially. Eli Lilly will only send to research laboratories. May be available through pharmaceutical suppliers in some countries.)

      The Lilly Research Laboratories
      Eli Lilly and Company
      Indianapolis, Indiana 46206

      Streptomycin sulfate
      Available from many chemical and pharmaceutical companies, for example:

      Calbiochem
      P.O. Box 12087
      San Diego, Calif. 92112

      Neomycin sulfate
      Available from many chemical companies. Commonly used in veterinary medicine, available in suspensions from veterinary suppliers.

      Penicillin G, potassium salt
      Available from many companies, eg. Calbiochem.

      Polymyxin B, sulfate
      Available from some chemical companies, eg. Calbiochem.

      Sodium Ampicillin (many trade names, eg. Polycillin)

Available through pharmaceutical suppliers, commonly
used in medicine as a substitute for Penicillin.

Possible sources of antibiotics for selective media

Anti-fungal, not strongly inhibitory to *Phytophthora*

Pimaricin (trade names Myprozine, Pimafucin)

Order under the name Pimaricin from:

Mycofarm-Delft
Royal Netherlands Fermentation Ind.
P.O. Box 1
Delft, Holland

(state clearly on order: for laboratory research on
fungi only, not for use with humans or animals).

Can also be obtained as Delvocid-Instant, a powder containing 50% active substance (= pimaricin) and 50%
lactose.

Distributors: Enzyme Development Corp.
              2, Penn Plaza
              New York, N.Y. 10001

Pentachloronitrobenzene, PCNB (trade names Terraclor, Brassicol) common agricultural fungicide

Olin Corporation
Agricultural Division
P.O. Box 991
Little Rock, Arkansas 72203

Mycostatin (Nystatin)

Available through various companies, for example:

1. Calbiochem
   P.O. Box 12087
   San Diego, Calif. 92112

2. Serva
   Heidelberg, Germany

(May be available through pharmaceutical suppliers in
some countries, since it has some medicinal uses.)

Benomyl (Methyl 1-(butyl carbamoyl)-2-benzimidazol) carba-
mate) (trade names Benlate, Tersan 1991). Common systemic fungicide, available through agricultural chemical suppliers.

E.I. du Pont de Nemours & Co. (Inc.)
Biochemicals Department
Wilmington, Delaware 19898

Adapted from information compiled by L. Klure, University of
California, Riverside, Ca.

C. Disadvandages of polyene antibiotics

1. The are not thermostable.

2. Some are inactivated by light.

3. Not readily available in some countries.

PCNB and Benomyl appear to be good alternatives to the use of
antibiotics.

PONCHET *et al.* 1972. Ann. Phytopathol. 4: 97-108.

FOLLIN, J.C. 1971. Cot. Lib. Tropicales. XXVI (4): 467.

D. Selective antibiotic media commonly used for the direct isolation of *Phytophthora* species from soil

Useful references:

TSAO, P.H. 1970. Selective media for isolation of pathogenic fungi. Annual
Review Phytopathology 8: 157-186.

I. Tsao's Medium

    1. Cornmeal agar . . . . 17 g/liter

       Pimaricin . . . . 10 ppm

       Vancomycin . . . . 200 ppm

       PCNB . . . . 100 ppm

    2. Dissolve or suspend the chemicals in sterile water as a concentrated stock solution and add to the molten cornmeal agar at 45-48°C in appropriate proportions before pouring the plates.

    3. When a water insoluble preparation of Pimaricin is used, first dissolve the antibiotic in a small amount of dimethylsulfoxide (DMSO) so that the final concentration of DMSO in the medium is 0.5% or less.

    4. This medium has been successfully used to isolate *Phytophthora* spp. from soil.

Note:

a) Pimaricin is sensitive to light. Isolation plates should be incubated in the dark to prevent loss of potency due to photoinactivation.

b) Prolonged storage of the medium, even in the dark and at a low temperature will result in loss of antibiotic activity.

TSAO, P.H. and G. OCANA. 1969. Nature (Lond.) 223: 636-638.

Possible substitutes for above:

Bacteria:

Streptomycin - try 100 ppm. *Phytophthora* spp. vary in sensitivity.

Penicillin and Polymyxin - try 50 ppm of each.

Neomycin - try approximately 70 ppm.

Fungi:

Nystatin (Mycostatin) use like Pimaricin (10-100 ppm). Dissolve in DMSO or ethyl alcohol, but keep the final concentration of solvents below 0.5%.

Benlate - try 20 ppm. Good for separating *Phytophthora* from *Fusarium*.

Unavailability of antibiotics

In some countries (e.g. Taiwan), Vancomycin is unavailable. In these situations, Ampicillin, which is readily available in most countries, can be substituted as follows:

```
V8 juice . . . . . . . . . . . . . . . . . . . . . 50 ml

Mycostatin . . . . . . . . . . . . . . . . . . . . 50 mg

Ampicillin . . . . . . . . . . . . . . . . . . . . 100 mg

PCNB . . . . . . . . . . . . . . . . . . . . . . . 10 mg

Agar . . . . . . . . . . . . . . . . . . . . . . . 20 g

Distilled water . . . . . . . . . . . . . . . . . 1000 ml
```

KO, W.H., M.S. CHANG, and H.J. SU. 1977. Proc. Amer. Phytopathol. Soc. 4: 74.

II. Schmitthenner's VYS-PCNB Medium

```
Distilled water . . . . . . . . . . . . . . . . . 1000 ml

V8 juice . . . . . . . . . . . . . . . . . . . . . 40.0 ml

CaCo₃ . . . . . . . . . . . . . . . . . . . . . . . 0.6 g

Bacto yeast extract . . . . . . . . . . . . . . . 0.20 g

Sucrose . . . . . . . . . . . . . . . . . . . . . 1.00 g

Cholesterol (2 ml N, N-dimethyl formamide soln.) . 0.01 g

Benlate (50% Benomyl) . . . . . . . . . . . . . . 0.02 g

Terrachlor (75% pentachloronitrobenzene) . . . . . 0.027 g
```

Neomycin sulfate . . . . 0.1 g

Chloroamphenicol . . . . 0.01 g

Bacto Agar . . . . 20.00 g

1. Autoclave the V-8 juice, CaCo$_3$ and distilled water for
   15 min at 121°C. Filter through Whatman #1 filter paper
   with a pad of celite 545 filter aid.

2. Add the sucrose, yeast extract, benlate, terrachlor, neomycin, and chloroamphenicol.

3. Disperse the cholesterol by shaking.

4. Add the agar and autoclave.

SCHMITTHENNER, A.F. 1973. Proc. First Woody Ornamental Disease Workshop 1:
94-110.

III. Sneh's Antibiotic Medium

1. KH$_2$PO$_4$ . . . . 1 g

   MgSO$_4$.7H$_2$O . . . . 0.5 g

   CaSO$_4$.2H$_2$O . . . . 0.1 g

   DL-threonine . . . . 1 g

   Thiamine-HCl . . . . 0.02 g

   Agar . . . . 20 g

   Sucrose . . . . 5 g

   Corn oil 0.1%

   PCNB . . . . 0.1 g

   Triton B-1956 . . . . 0.1 ml

   ☆Pimaricin (Sodium salt) . . . . 2 ppm

   ☆Mycostatin (3920 units/mg) . . . .
   . . . . 10 ppm (dissolved in 0.5% DMSO)

   ☆Chloramphenicol . . . . 60 ppm

   ☆Penicillin "G" (sodium) (1000 u/g) . . . . 60 ppm

   ☆Polymixin B sulfate (6300 u/mg) . . . . 60 ppm

☆Before pouring into plates, add these antibacterial and antifungal components to the molten medium.

2. Pipette 1-ml aliquots of 10^{-1} soil dilutions on the solidified agar medium.

3. After incubation of 24 hrs at $25^{\circ}C$, wash the soil from the plates and add 0.1-ml aliquots each of fresh solutions of rose bengal and pimaricin to give a final concentration in the medium of 10 and 80 ppm, respectively.

4. The medium is satisfactory for the isolation of *P. cactorum*, *P. nicotianae* var. *parasitica*, *P. cinnamomi* (without rose bengal), *P. drechsleri* (when NH_4NO_3 is substituted for DL-threonine). Not suitable for *P. cryptogea*, *P. cambivora* and *P. megasperma* var. *sojae*.

SNEH, B. 1972. Can. J. Microbiol. 18: 1389-1392.

IV. Isolation of *Phytophthora* in tropical areas

Lima Bean Agar (Difco) . . . . 20 g/l

Polymixin . . . . 50 ppm

Penicillin . . . . 50 ppm

Benomyl . . . . . 50 mg/liter (50% active ingredients)

Notes:

Benomyl is a good replacement for pimaricin because of:

a)   Its stability.

b)   Its availability in most countries.

c)   Its ability to withstand autoclaving.

d)   Its ability to reduce the incidence of fungal saprophytes commonly isolated on other antibiotic media.

FOLLIN, J.C. 1971. Cot. Lib. Tropicales. Vol. XXVI (4): 467.

E. The presence of *Pythium* in *Phytophthora* isolations

All the above antibiotic media result in the isolation of *Py-thium* as well as *Phytophthora*. This can often prove to be a problem since *Phytophthora* is easily masked or overgrown by the more rapidly growing *Pythium* spp.

A chemical, Hymexazol (see below), is now available that inhibits *Pythium* and allows the continued growth of *Phytophthora*. Hymexazol is thus a major breakthrough in increasing the selectivity of antibiotic media for *Phytophthora* isolations.

Ia.) Hymexazol (HMI) medium for inhibition of *Pythium* spp.

    1. Potato dextrose agar (PDA) 1%

       Benomyl (Benlate 50% active) . . . . 10 μg/ml

       Nystatin (Mycostatin 2000 units/mg). . . . . . 25 μg/ml

       PCNB . . . . 25 μg/ml

       Rifampicin (Rifampin) . . . . 10 μg/ml

       Ampicillin (Viccillin) . . . . 5000 μg/ml

       HMI-3-hydroxy-5-methylisoxazole-(Trachigaren). . . . . .

       . . . . 25-50 μg/ml

    2. Dissolve or finely suspend each of the above antibiotics in 80% ethanol as a concentrated stock solution, and mix in appropriate proportion with sterile distilled water to give a ten-fold concentration of that to be used in the final medium.

    3. Add the chemicals to the petri plate containing the soil sample, being careful not to contact the soil with the chemicals.

    4. Add molten PDA agar at approx. 45°C to the petri plate and shake to disperse all ingredients evenly in the plate.

Notes:

a) This method can be used to determine propagule densities of
    Phytophthora in the soil since the medium does not inhibit
    germination and colony establishment of oospores and chlamy-
    dospores.

b) HMI derivatives, 3-hydroxy-5-ethyl- and 3-hydroxy-5-n pro-
    pylisoxazoles also show differential toxicity toward *Phy-
    tophthora* and *Pythium*.

c) Make sure that the final concentration of ethanol used to
    dissolve the chemicals does not exceed 0.5%.

d) HMI (Tachigaren) is obtainable from Sankyo Co. Yasu, Shiga-
    Ken, Japan. Ampicillin (Viccillin) is obtainable from Meiji
    Seika, Tokyo, Japan.

MASAGO, H., M. YOSHIKAWA, M. FUKADA, and N. NAKANISHI. 1977. Phytopathology
67: 425-428.

I(b) Hymexazol medium for inhibition of *Mortierella* and *Pythium*

1. Sterilize 17 g Difco cornmeal agar in 1000 ml distilled
   water, and cool to 45-48°C.

2. To the molten cornmeal agar add the following chemicals:

Note:

These chemicals are first dissolved or suspended in sterile
distilled water as stock solutions and added to the molten cornmeal agar in appropriate proportions before pouring the plates.

Hymexazol (3-hydroxy-5-methylisoxazole) . . . . . 50 μg/ml

Vancomycin . . . . 200 mg

PCNB (Terraclor, 75% wettable powder) . . . . 100 mg

Pimaricin (Pimafucin, 90.5%) . . . . 10 mg

3. This medium effectively inhibits *Pythium* spp. and *Mor-*

tierella spp., thus greatly facilitating the detection
and enumeration of *Phytophthora* spp. in soil dilution
plates.

TSAO, P.H. and S.O. GUY. 1977. Phytopathology 67: 796-801.

Section 2

Isolation Techniques

2D: Antibiotic media for isolation of specific

Phytophthora species

I. Isolation of *P. cactorum* from soil

1. Sucrose . . . . 20 g

   DL-threonine . . . . 1 g

   KH_2PO_4 . . . . 1 g

   $MgSO_4.7H_2O$ . . . . 0.1 g

   $CaSO_4.2H_2O$ . . . . 0.1 g

   Thiamine-HCL . . . . 0.02 g

   Agar . . . . 20 g

   Distilled water . . . . 1000 ml

2. Add 10 ppm Benomyl dissolved in 0.5% dimethyl sulfoxide (DMSO), to the molten medium to suppress non-pythiaceous fungi.

3. Add 200 ppm sodium taurocholate to increase the sensitivity for the recovery of *P. cactorum*.

4. Place soil samples directly on the agar medium.

5. Pimaricin (80 ppm, wt. pimaricin/vol. of agar), may be added before or after washing the soil from the plates.

6. Germinating cysts and sporangia can be recognized 16-24 hrs. after adding the soil dilutions to the agar medium.

Note:

Numbers of germinated oospores are greater if the soils are incubated on the medium for 48 hrs.

SNEH, B. 1974. Can. J. Botany 52: 1777-1778.

McINTOSH, D.L. 1975. Can. J. Botany 53: 1444-1445.

II. Isolation of *P. cinnamomi* from soil

1. Make up the following medium:

   $NaNO_3$, 2.0 g; $MgSO_4$, $7H_2O$ .05 g; $FeSO_4$, 1.0 g; KCl, 0.5 g;

Sucrose 30 g; yeast extract, 0.5 g; Agar, 15 g; Streptomycin sulfate, 50 ppm; Rose bengal, 60 ppm; PCNB, 1000
ppm; Mycostatin, 100,000 units, and distilled water,
1000 ml.

2. Combine all ingredients except mycostatin and PCNB prior
to autoclaving.

3. Stocks of the medium without these two ingredients can
be stored in a dark cabinet at room temperature.

4. When plates are needed, melt the medium, cool to 40-42°C,
add PCNB and adjust pH to 4.8 with lactic acid.

5. Add an aqueous suspension of mycostatin just prior to
pouring.

6. Use the plates within 6 hrs.

7. Place soil dilutions of 1:2; 1:4; and 1:10 on the plates
and incubate at room temperature 72-96 hours. After 120
hours, other organisms develop rapidly.

8. The pH of the medium is critical. Maximum recovery of *P.
cinnamomi* is obtained only at an initial pH of 4.5-5.0.

9. Recovery of *P. cinnamomi* is greatly reduced when soil is
plated on media with less than 2% agar.

10. Using 4% agar, bacterial colonies are restricted, while
growth of the fungus is considerably enhanced.

11. The authors state that this medium can be used to estimate population densities of *P. cinnamomi* in a given soil

12. Incubation of the plates at high temperature in light, or
for a long period of time, results in loss of inhibition
toward many contaminating organisms.

Note:

P. cactorum and *P. parasitica* also grow on this medium, but *P.
cinnamomi* can easily be distinguished by the presence of characteristic hyphal swellings that form on the mycelium (see Morphology section).

HENDRIX, F.F. and E.G. KUHLMAN. 1965. Phytopathology 55: 1184-1187.

III. Isolation of *P. infestans* from blighted leaves

1. Frozen peas (Birds Eye) . . . . 160 g

   Davis agar . . . . 20 g

   (w/v) "Vancocin" 1510 units/mg . . . . 1000 ppm

   (v/v) "Pimafucin" (Pimaricin) . . . . 2000 ppm

   (w/v) Chloramphenicol . . . . 50 ppm

   Water . . . . 1000 ml
      The pimaricin and vancomycin were added to the
      strained pea liquor before autoclaving at 15 lb/15
      min; chloramphenicol was added as the solid after
      autoclaving.
2. Pieces of leaf adjacent to the blight region are placed
   directly on the agar after surface sterilization.
3. The fungus can be subcultured after 4-6 days incubation
   at 21°C.

HOLLOMON, D.W. 1965. Plant Pathology 14: 34-35.

IV. Isolation of *P. megasperma* var. *sojae* on soil dilution
    plates

1. Supplement Difco cornmeal agar with pimaricin (2 ppm),
   penicillin "G" (80 units/ml), polymixin B (370 units/ml)
   and PCNB (pentachloronitrobenzene) (10 ppm).
2. Adjust the pH to 4.6 with lactic acid.
3. The concentration of pimaricin is critical. Direct isolation from soil is prevented at doses of pimaricin higher
   than 2 ppm even though *Phytophthora* transferred from pure
   culture grows at 100 ppm.
4. Dilute screened soil 1:1000 in 1% water agar.
5. Seed each plate with 0.5 ml of the soil mixture.

Notes:

a) Mixing liquid agar with soil in petri plates yields more
    colonies than distributing the soil on solid agar.

b) This method is successful for isolation from the soybean
    rhizosphere, but not from soil that has not had recently
    infected plants growing in it.

HAAS, J.H. 1964. Phytopathology 54: 894.

V(a) Flowers and Hendrix's medium for isolation of *P. nicotianae*
     var. *parasitica*

1. Sucrose . . . . 30 g

   NaNO$_3$ . . . . 2 g

   MgSO$_4$.7H$_2$O . . . . 0.5 g

   KH$_2$PO$_4$ . . . . 1 g

   Yeast extract . . . . 0.5 g
   Thiamine-HCl . . . . 2 mg

   Gallic acid . . . . 425 mg

   Rose bengal . . . . 0.5 mg

   PCNB . . . . 80,000 units

   Nystatin . . . . 100,000 units

   Distilled water . . . . 1 liter

   Agar . . . . 20 g

        The pH of the medium is 4-5. Add PCNB, Penicillin
        "G" and Nystatin after autoclaving the rest of the
        medium

2. After pouring the plates, chill at 3°C before use.

3. Seed chilled plates with 1 ml of 1:50 soil suspension
   and incubate at 24°C for 26 hrs in the dark.

4. Wash soil off the agar with slowly running tap water before observing colonies.

5. Poured plates can be stored in the cold and dark for
   approx. 2 weeks.

Note:

Not successful for isolating *P. megasperma* due to large numbers
of *Pythium* spp. in alfalfa soil samples (cf. Marks & Mitchell,
1970 Phytopathology 60: 1687-1690.

FLOWERS, R.A. and J.W. HENDRIX. 1969. Phytopathology 59: 725.

V(b) Ponchet et al's medium for the isolation of *P. nicotianae*
     var. *parasitica*

1. Malt agar . . . . 10 g

   Penicillin "G" . . . . 250 ppm (400,000 IU)

   Polymixin B . . . . 250 ppm (200,000 IU)

   Benomyl . . . . 15 ppm

   PCNB . . . . 100 ppm

   Distilled water . . . . 1000 ml

2. After autoclaving, inoculate the medium (after cooling
   to 45°C), with soil suspensions and pour into petri
   plates.

Note:

Pythium and *Mortierella* are not inhibited by this medium and
may interfere with *Phytophthora* isolations.

PONCHET, J., P. RICCI, C. ANDREOLI, and G. AUGE. 1972. Annales Phytopathol.
: 97-108.

Section 3:

Culture Media

3A Natural media for

growth and sporulation

Natural media for the growth and sexual reproduction of *Phytoph-thora* spp.

Ia) V-8 Juice (Non-clear)

1. Add 800 ml of deionised distilled water to 200 ml of V-8 juice (Campbell Soup Co.). This will give a 1 to 4 dilu-tion of the V-8 juice.

2. Add 2 g $CaCO_3$ to the diluted juice and mix on a stir plate.

3. Add 15 g agar and autoclave.

4. Shake flask often while dispensing to ensure good disper-sion of $CaCO_3$.

Ib) V-8 Juice (Clarified)

1. Add 5 g $CaCO_3$ to one 12 oz can of V-8 juice (354 ml), and mix on a stir plate.

2. Centrifuge for 20 min at 4000 rpm.

3. Decant the clear solution carefully into a flask.

4. For broth media, dilute the supernatant 1 to 4 with deionised distilled water, autoclave and dispense.

5. For agar media, add 15 g agar per liter (1.5%) to the broth, autoclave and dispense.

Notes:

a) This medium has proved to be the most popular and success-ful medium for the growth and sporulation of most *Phytoph-thora* spp.

b) For the detailed chemical analysis of V-8 juice, see J.W. Hylin *et al*. Mycologia 62: 200-202 (1970).

II. Cleared lima bean agar

285 g of frozen lima beans (not blended) are autoclaved at
121°C for 15 min. in 1000 ml distilled water. The beans are removed by straining through a cotton cloth, 17 g agar are added
and the volume adjusted to 1000 ml before reautoclaving at
121°C for 15 min The agar is dispensed aseptically, 15-20 ml/
sterile plate.

The following can be substituted for lima beans:

    1. Corn meal agar: 60 g of Quaker yellow cornmeal, or 20 g
                               Difco cornmeal.

    2. Oat meal agar: 50-60 g of Quaker oatmeal, or 20 g Difco
                               oatmeal.

    3. Hemp seed agar: 50 g of hemp seed.

    4. Soybean agar: 100 g of soybean seed.

The hemp and soybean seeds are soaked overnight before the agar
medium is prepared following the procedure for cleared lima
bean agar.

SAVAGE *et al*. 1968. Phytopathology 58: 1004-1021.

III. Oat grain medium

    1. Soak non-hulled grains in distilled water 2-3 hours.

    2. Pour off free water and autoclave 30 min. at 15 psi.

    3. Top flasks with aluminum foil.

    4. Will keep at 20°C for 2-3 months.

ROYLE, D.J. and C.J. HICKMAN. 1964. Can. J. Botany 42: 311-318.

IV. Lima-bean/V-8 juice medium

    1. Autoclave 80 g frozen lima beans in 200 ml DD water for
       8-10 min.

2. Strain through cheese cloth and discard the lima beans.

3. Add the supernatant to 100 ml clarified V-8 juice (see page 75).

4. Bring the volume of the lima bean/V-8 juice to 1000 ml with DD water.

5. Adjust to pH 6.0 with KOH or HCl.

6. Add 14 g Difco agar and autoclave (15 min. at 15 psi.).

RIBEIRO, O.K. Dept. of Plant Pathology. Univ. of California, Riverside.

V. Modified Carrot agar

1. Slice 200 g fresh carrots and place in a blender cup.

2. Add 500 ml distilled water.

3. Blend for 40 sec. at high speed.

4. Filter through four layers of cheesecloth, and squeeze out juice from the residue.

5. Add distilled water to bring volume up to one liter (1000 ml).

6. Add 1.5% agar and autoclave.

Note:

This medium is useful for oospore production of many *Phytophthora* spp., particularly *P. palmivora*.

KAOSIRI, T. Dept. of Plant Pathology. University of California, Riverside.

VI. Red kidney bean meal agar; *P. fragariae*

1. Prepare 28 g red kidney bean meal with 13.6 g agar and 800 ml water in a 2 liter flask.

2. Add one drop of Antifoam, FG-10 (Food Grade) emulsion

(Dow Corning Corp., Midland, Michigan), per 100 ml   of
medium to reduce tendency of the medium to boil over
when autoclaved.

Note:

Not all races of *P. fragariae* will produce oospores on this medium.

MAAS, J.L. 1972. Mycopathol. Mycologia Appl. 48: 323-324.

VII. Rape seed extract agar

100 grams of rape seed are thoroughly washed in water and boiled in 1 liter of distilled water, for about 30 min. The extract
is then filtered through cotton and adjusted to the original
volume. Twenty grams of agar are added to the extract and the
mixture autoclaved at 15 psi. for 15 min. The initial pH of the
medium is adjusted to approx. 7.00.

SATOUR, M.M. 1967. Mycologia 59: 161-166.

VIII. Incorporation of charcoal in *Phytophthora* growth media

1. Add 1 to 10% of activated charcoal to any growth medium
   of *Phytophthora*.
2. The addition of this absorbant improves growth of the
   fungus and does not prevent the production of sporangia
   and oospores.

PARMENTIER, G. 1970. Parasitica 26: 31-40.

Section 3B:

Chemically defined media for

growth and sporulation

A. The following chemicals are essential for growth of *Phytoph-thora* in axenic culture:

1. A carbon source (glucose, sucrose, etc.). *See also:*
   J.W. Hendrix and J.L. Apple. 1964. Fats and fatty acid
   derivatives as growth stimulants and carbon sources for
   Phytophthora parasitica var. *nicotianae*. Phytopathology
   54: 987-994.

2. A nitrogen source. The following are good sources of N:
   L-α-alanine, L-asparagine, L-aspartic acid, L-glutamine,
   L-glutamic acid, glycine, L-histidine, L-proline, L-
   serine, L-threonine and L-tyrosine. Poor sources of N
   are: L-hydroxyproline, L-isoleucine, L-leucine, L-lysine,
   L-methionine and L-tryptophan. (Leal *et al*. Can. J.
   Microbiol. 17: 1319-1325, 1971; Ribeiro, O.K. Unpublished, 1975). *Caution:* Ammonia accumulation from use of
   organic N can lead to decrease in growth and toxicity to
   sporangia and oogonia (Leal *et al*. Mycologia 63: 1041-
   1056, 1970).

3. Fe^{3+} (Cuppett & Lilly. Mycologia 65: 67-77, 1973).

4. Ca^{2+} (Erwin, D.C. Mycologia 60: 1112-1116, 1968; Elliott,
   C.G. Trans. Brit. Mycol. Soc. 58: 169-172, 1972). An
   absolute requirement for Ca^{2+} has been reported for *P.
   fragariae* by Davis, 1969 (Trans. Brit. Mycol. Soc. 42:
   193-200). Hendrix *et al*. 1969 (Phytopathology 59: 1620-
   1624), have shown that Ca^{2+} concentration for maximum
   growth is lower for cultures grown in a medium containing cholesterol than for cultures devoid of cholesterol.

5. Zn^{2+} has been reported by some investigators to be essential for optimum growth of *Phytophthora*.

Notes:

a) Since physiological differences are known between isolates
    of the same species (Kelley, W.D. Can. J. Microbiol. 21:
    1548-1552, 1975; Ribeiro, O.K. Unpublished, 1975; Zentmyer
    et al. Phytopathology 66: 932-986, 1976), a single medium

may not be sufficient for the optimum growth of several different isolates.

b) Growth of *P. megasperma* in axenic culture requires the presence of sterols (Erwin *et al.* Phytopathology 58: 1049, 1968) All the other species in the genus do not require sterols for growth, but growth is stimulated in the presence of sterols. Sterols however, are an absolute requirement for sexual reproduction of all species of *Phytophthora* on chemically defined media (Hendrix, J. Annual Rev. Phytopathol. 8: 111-130).

c) Chelators: Conflicting reports on the toxicity of chelators to fungi have been reported by various investigators. Erwin and Katznelson, 1961 (Can. J. Microbiol. 7: 15-25), found that ethylene diamine tetraacetic acid (EDTA) at 25 or 50 mg/l, is toxic in the presence of minor elements to *P. cryptogea*. Hendrix *et al.* 1969 (Phytopathology 59: 1620-1624), however, found EDTA to be toxic only in the absence of minor elements. They also found Ca^{2+} to be ineffective in reducing EDTA toxicity, while Erwin and Katznelson (1961), found that Ca^{2+} reduced EDTA toxicity. Ribeiro, 1975 (Mycologia 67: 1012-1019), found no evidence of EDTA toxicity when incorporated with Fe^{3+}.

d) Effect of O_2 and CO_2: Growth of *Phytophthora* on solid media is reduced at O_2 levels below 5% and stimulated at 5% CO_2 when the O_2 level is 1%. Above 5% CO_2, growth is reduced. At 15% CO_2 and 20% O_2, growth in liquid or solid media is reduced (Mitchell and Zentmyer. Phytopathology 61: 787-791, 1971).

Useful References:

(1) COCHRANE, V.W. 1958. Physiology of Fungi. John Wiley & Sons, Inc. New York, 524 pp.

(2) LILLY, V.G. 1965. The chemical environment for fungal growth. 1. Media, Macro- and Microelements, pages 465-478. *In*, The Fungi, Vol. 1 (G.C. Ainsworth & A.S. Sussman, eds.).

(3) LILLY, V.G. and H.L. BARNETT. 1951. Physiology of the Fungi. McGraw-Hill Book Co. Inc. New York, 464 pp.

B. Sterols: Methods for dispersion into solutions

Sterols are difficult to get into solution since they are insoluble in water.

The following methods have been reported for dissolving sterols used in chemically defined media for growth and sporulation of *Phytophthora*.

1. Dissolve the crystalline sterol in dichloromethane (CH_2Cl_2), on a 1:1 basis e.g., 20 mg sterol/20 ml CH_2Cl_2. Dichloromethane evaporates when the medium is autoclaved.

2. Dissolve 0.1 g of the crystalline material in 1 ml of hot polyoxyethylene sorbitan mono-oleate (Tween 80). Add to medium.

3. Dissolve sterol in dimethylformamide (DMF). Use 2.0 mg of cholesterol/ml DMF.

4. Dissolve sterol in dimethyl sulfoxide (DMSO).

5. Dissolve in ethanol. Note: Ethanol can be toxic to the fungus at concentrations of 1% and above.

6. Dissolve in 1-2 ml of ether. Ether evaporates on autoclaving.

A summary of some chemically defined media used in the growth and reproduction of *Phytophthora* are given in Table 2.

Note:

It is advisable to autoclave the carbon source e.g., glucose, sucrose, etc. separately from the rest of the medium and combine the sterilized ingredients, just prior to dispensing of the medium.

Table 2: Comparison of Defined Media for Growth & Reproduction of *Phytophthora*

Authority	Sucrose (g/l)	Glucose (g/l)	L-Aspara-gine (g/l)	Thiamine (mg/l)	KH_2PO_4 (g/l)	K_2HPO_4 (g/l)	$MgSO_4$. $7H_2O$ (g/l)	Beta Sitos (mg/l)
Lopatecki 1956	50.0		4.0	20.0	1.0		0.1	
Hodgson 1958	5.0		1.0	200.0			1.0	
Davis 1959		5.0		1.0	0.5		0.25	
Erwin & Katznelson 1961	15.0		2.0	1.0	0.43	0.30	0.1	
Leal et al. 1964	30.0		2.0	1.0	0.6		0.3	
Cameron & Milbrath 1965	25.0		5.0	0.02	0.75	0.75	0.5	
Bartnicki-Garcia 1966		20.0		1.0	2.0		0.5	
Cameron 1966	5.0		5.0	0.012	0.75	0.75	0.5	
McCormick & Erwin 1967				1.0	1.9	1.04	0.1	
Leal & Gomez-Miranda 1964		5.0	2.0	1.0		0.5	0.5	10.0
Leal et al. 1967		5.0	0.1	1.0	0.5		0.5	10.0
Leal et al. 1967		5.0		1.0	0.5		0.5	10.0
Mircetich et al. 1969	15.0		2.0	1.0	0.43	0.30	0.1	
Castro et al 1969		20.0		1.0	2.0		0.5	
Timmer et al. 1969		20.0		1.0	0.5		0.2	30.0
Mitchell 1970	15.0			1.0	0.5		0.2	30.0
Mitchell 1970		5.0		1.0	0.5		0.2	30.0
Erwin & Partridge 1970	15.0		2.0	1.0	1.9053	1.0451	0.2	30.0
Huguenin & Boccas 1971		10.0		2.0	0.8	0.1	0.5	20.0
Elliott 1972	10.0		1.0	1.0	0.5		0.25	
Cuppett 1973		5.0	0.5		1.0		0.5	10.0
Ribeiro 1974		4.5	0.1	1.0	1.0		0.5	30.0

C.   Chemically defined media: Individual recipes

Phytophthora spp. are not uniform in their nutritional requirements for growth and reproduction. Some species will grow on
relatively simple substrates containing a few salt solutions
and a carbon and nitrogen source, while other species require
a high degree of nutritional complexity for successful growth
(Table 3). Even isolates of the same species differ in their
nutritional requirements.

The following media have been reported to support good growth
and/or reproduction of *Phytophthora* spp.

$Na_3MoO_4\cdot 2H_2O$ (mg/l)	$FeSO_4\cdot 7H_2O$ (mg/l)	$ZnSO_4\cdot 7H_2O$ (mg/l)	$CuSO_4\cdot 5H_2O$ (mg/l)	$MnSO_4\cdot H_2O$ (mg/l)	$MnSO_4\cdot 4H_2O$ (mg/l)	$(NH_4)_6MoO_{24}\cdot 4H_2O$ (mg/l)	Casein Hydrolysate (g/l)	Infrequently used chemicals
0.02	1.0	1.0	0.02	0.02				$CaSO_4$ (0.1g); $FeCl_3\cdot 6H_2O$ (1.0mg); K_3PO_4 (1.0g)
0.02	1.0	0.16	0.04	0.02				L-Tyrosine (0.5g)
	0.02	1.0	0.02					$MnCl_2\cdot 7H_2O$ (0.02mg)
0.02	0.5	0.5	0.02					$MnCl_2\cdot 4H_2O$ (0.02mg); $Ca(NO_3)_2$ (0.05g)
								NH_4NO_3, 1.0g; Citric Acid, 1.0g
	1.0	1.8	0.4	0.3		0.3		$NaNO_3$, 0.5g
		1.46	0.79		0.405	0.3		$NaCl$, 0.1g; H_3BO_3, 0.285mg; MoO_3, 0.175g, $FeC_6H_5O_7\cdot 3H_2O$ 5.35mg
0.02	1.0	1.0	0.02	0.02			6.0	
0.02	0.5	0.5	0.02		0.02			
		0.2	0.02	0.05				$Fe(NO_3)_3\cdot 9H_2O$, 0.7mg
		0.2	0.02	0.05				KNO_3, 3.05g
0.02	1.0	1.0	0.02		0.02			
	1.0	1.8	0.4			0.3	2.0	Dl-Cittruline, D-L Serine, Inositol, Biotin, $B_2 B_6$, $NaNO_3$, 0.5g, Citric Acid, 1.4mg
	1.0	1.8	0.4			0.3		$NaNO_3$ (0.5g), Citric Acid, 1.4mg; Yeast
0.02	1.0	1.0	0.02	0.02				2.0g
0.02	1.0	1.0	0.02	0.02				2-(N-morpholino ethanesulphonic acid, 5.3g
	1.0					0.02		KNO_3, 0.154g
0.52	1.0	1.0	0.04			0.04		$MnCl_2\cdot 7H_2O$, 0.04mg; K_2SO_4, 0.5g; KNO_3, 1.0g; $TiSO_4\cdot 5H_2O$, 0.04mg
0.05		4.4	0.393		0.072			
		0.2	0.02	0.05		0.1		Ascorbic Acid, 200mg; Ferric iron, 1ppm
0.01		0.2	0.02		0.05			$FeCl_3\cdot 6H_2O$, 0.1mg in EDTA & KOH, $Na_2B_4O_7$, 0.088mg, KNO_3, 0.148g

I. Hohl's medium P-1 for growth

Chemicals	g/l
$NaNO_3$	3.0
K_2HPO_4	1.0
$MgSO_4\cdot 7H_2O$	0.5
KCl	0.5
$CaCl_2\cdot 2H_2O$	0.1
$FeSO_4\cdot 7H_2O$	0.01
$FeCl_3$	0.001
$ZnCl_2$	0.001

Hohl's grouping of species according to their general growth requirements are shown in Table 3.

Table 3: Grouping of some *Phytophthora* species according to their nutritional requirements for vegetative growth

☆Group 1	Group 2	Group 3	Group 4
Inorganic N	Inorganic N+ Lecithin	Organic N+ Lecithin	Organic N+ Lecithin + Organic factors
P. capsici	*P. cactorum*	*P. erythroseptica*	*P. fragariae*
P. cryptogea	*P. cambivora*	*P. infestans*	*P. infestans*
P. vesicula	*P. cinnamomi*	*P. megasperma* var.	
	P. citricola	*megasperma*	
	P. citropthora	*P. syringae*	
	P. lateralis		
	P. nicotianae var.		
	nicotianae		
	P. palmivora		

☆Group 1 = Species having simple nutritional requirements.

Group 4 = Species having complex nutritional requirements.

P. infestans isolates differ in their nutritional requirements.

Inorganic N = $NaNO_3$ (1.5g/liter).

Organic N = Asparagine.H_2O (1.0g/liter).

HOHL, H.R. 1975. Phytopathol. Z. 84: 18-33.

EDTA	0.01
Thiamine-HCl	0.0005
Guanine	0.05
Asparagine.H_2O	0.130
Histidine, HCl	0.190
Serine	0.05

Chemicals	g/l
Leucine	0.065
Cysteine	0.06
Sucrose	30.0
Agar	15.0
Lecithin	0.2
Sterol (cholesterol or stigmasterol)	0.05

If lecithin does not dissolve easily, homogenize with 5-10 ml of the medium in a Virtis high speed homogenizer, or in a Braun-Sonic ultrasonic homogenizer.

Sterols can be dissolved in 1-2 ml of ether.

Grind purines and pyrimidines with mortar and pestle before adding to the medium.

The following species have been reported to make good growth on this medium: *P. cactorum, P. cambivora, P. capsici, P. cinnamomi, P. citricola, P. citrophthora, P. cryptogea, P. erythroseptica, P. fragariae, P. infestans, P. lateralis, P. megasperma, P. nicotianae* var. *nicotianae, P. palmivora, P. syringae,* and *P. vesicula.*

Notes:

1. Addition of lecithin gave increases in linear growth rate from 60-300%.

2. *P. infestans* was found to have an absolute requirement for a purine base.

3. Most species will grow without the amino acid mixture, provided nitrate and/or asparagine are present in the medium.

II. Erwin's synthetic medium (ESM) for growth

A.   15 g. sucrose in 200 ml total vol DD H_2O

    2.00 g   L-Asparagine
    200 mg $MgSO_4 \cdot 7H_2O$
    1   mg $FeSO_4 \cdot 7H_2O$
    10 mg $CaCl_2 \cdot 2H_2O$
    1   mg   Thiamine HCl
    1   ml   Minor Element Solution:

B.   4.3983 g $ZnSO_4 \cdot 7H_2O$ = 1000 ppm Zn^{2+}
    0.0786 g $CuSO_4 \cdot 5H_2O$ = 20   ppm Cu^{2+}
    0.0527 g $NaMoO_4 \cdot 2H_2O$ = 20   ppm Mo^{6+}
    0.0204 g $MnCl_4 \cdot 2H_2O$

in 600 ml total volume of DD H_2O

C.   1.0451 g K_2HPO_4
    1.9053 g KH_2PO_4

in 200 ml total volume DD H_2O

Autoclave 12 min. at 15 psi. Bring to volume with sterile DD H_2O. Combine, when cool, A, B, and C.

If sterols are to be added, first dissolve sterol in ca. 50 ml CH_2Cl_2 (Dichloromethane) (30 mg/l is the normal sterol concentration), and add to part A before autoclaving.

pH should be between 6.55 and 6.60 after autoclaving. pH of L-asparagine solution may be adjusted with NaOH.

Note:

See following table for concentrations of ingredients to be used in making up various quantities of the above medium.

Table 4: Concentrations of ingredients required for various volumes of Erwin's synthetic medium

	500 ml	1000 ml	1500 ml	2000 ml	2500 ml	3000 ml
ucrose	7.5 g	15 g	22.5 g	30 g	37.5 g	45 g
	100 ml	200 ml	300 ml	400 ml	500 ml	600 ml total vol. DD H_2O
-Asparagine	1.0 g	2.0 g	3.0 g	4.0 g	5.0 g	6.0 g
$gSO_4 \cdot 7H_2O$	100 mg	200 mg	300 mg	400 mg	500 mg	600 mg
$eSO_4 \cdot 7H_2O$	.5 mg	1 mg	1.5 mg	2 mg	2.5 mg	3 mg
$aCl_2 2H_2O$	5 mg	10 mg	15 mg	20 mg	25 mg	30 mg
hiamine HCl	.5 mg	1 mg	1.5 mg	2.0 mg	2.5 mg	3 mg
inor Elements	.5 ml	1.0 ml	1.5 ml	2.0 ml	2.5 ml	3.0 ml
	300 ml	600 ml	900 ml	1200 ml	1500 ml	1800 ml total vol. DD H_2O.
$_2HPO_4$	0.52255 g	1.0451 g	1.56765 g	2.0902 g	2.61275 g	3.1353 g
H_2PO_4	0.95265 g	1.9053 g	2.85795 g	3.8106 g	4.76325 g	5.7159 g
	100 ml	200 ml	300 ml	400 ml	500 ml	600 ml
-sitosterol	15 mg	30 mg	45 mg	60 mg	75 mg	90 mg

.E. PARTRIDGE and D.C. ERWIN. 1969. University of California, Riverside.

III. Bartnicki-Garcia's liquid medium for growth

	per liter
Glucose	20 g
L-asparagine	2 g
KH_2PO_4	3 g
$CaCl_2$	3.4 mg
$FeSO_4 \cdot 7H_2O$	1.0 mg
Citric acid	1.4 mg
$ZnSO_4 \cdot 7H_2O$	1.8 mg
$MnSO_4 \cdot H_2O$	0.3 mg
$CuSO_4 \cdot 5H_2O$	0.4 mg
$(NH_4)_6Mo_7O_{24} \cdot 4H_2O$	0.3 mg
$MgSO_4 \cdot 7H_2O$	0.5 g
Thiamine HCl	1.0 mg
Distilled water	1000 ml

Autoclave glucose in concentrated solution separately and add
to the rest of the ingredients which should be previously adjusted to pH 6.0 with KOH and autoclaved for 20 min. at 121°C.

This medium has proved to be useful for the routine growth of several *Phytophthora* spp. It is not a good medium for sexual reproduction.

BARTNICKI-GARCIA, S. 1966. J. Gen. Microbiol. 42: 57-69.

IV. Lopatecki and Newton's medium:

Composition of nutrient solutions, growth therein, and pH at the end of the incubation period

Solution	1	2	3	4	5	6	7	8	9
Composition, gm/liter									
Dextrose	50	50	50	50	50	50	50	50	50
KH_2PO_4	1	1	1	1	1	1	1	1	1
Thiamine	0.02	0.02	0.02	0.02	0.02	0.02	0.02	0.02	0.02
KNO_3	4	4	4	4	--	--	--	--	--
$MgHPO_4$	--	0.1	--	--	--	--	--	--	--
$MgSO_4$	--	--	0.1	0.1	0.1	0.1	0.1	0.1	0.1
$CaSO_4$	--	--	--	0.1	0.1	0.1	0.1	0.1	0.1
$(NH_4)_2HPO_4$	--	--	--	--	4	--	--	--	--
Asparagine	--	--	--	--	--	4	4	4	--
dl-Alanine	--	--	--	--	--	--	--	--	4
$FeSO_4$	--	--	--	--	--	--	0.001	0.001	0.001
Trace minerals*	--	--	--	--	--	--	--	1 ml	1 ml
Growth, mgm.									
P. parasitic	Nil	52	63	112	49	118	210	245	186
P. cactorum	Nil	4	7	2	35	46	111	150	146
P. megasperma	Nil	1	10	65	5	113	178	269	239
P. erythroseptica	Nil	67	73	115	39	131	210	233	175
pH									
P. parasitica	--	5.6	5.8	5.6	3.2	6.1	6.0	5.9	5.9
P. cactorum	--	4.7	4.8	3.7	3.3	6.0	6.1	6.1	6.0
P. megasperma	--	4.2	4.3	4.6	3.6	5.9	5.8	6.0	6.0
P. erythroseptica	--	5.4	5.4	5.6	2.8	6.2	5.8	6.2	6.2

*1 ml of the trace mineral solution adds 1 ppm zinc and 0.02 ppm molybdenum, copper, and manganese.

LOPATECKI, L.E. and W. NEWTON. 1956. Can J. Botany. 34: 751-757. Reproduced by permission of the National Res. Council of Canada, Copyright, 1956.

V. Rapid growth of several species of *Phytophthora*

Medium:

D-glucose . . . . 10 g

$(NH_4)_2SO_4$ . . . . 2 g

Fumaric acid . . . . 2 g

$MgSO_4.7H_2O$ . . . . 0.5 g

KH_2PO_4 . . . . 1 g

Fe^{+++} $(FeNH_4[SO_4]_2.12H_2O)$ . . . . 0.2 mg

Zn^{++} $(ZnSO_4.7H_2O)$ . . . . 0.2 mg

Mn^{++} $(MnSO_4.H_2O)$ . . . . 0.1 mg

Thiamine hydrochloride . . . . 100 μg

β-sitosterol . . . . 2 mg

Distilled water . . . . 1000 ml

pH adjusted to 6,0

1. Dispense 25 ml portions into 250 ml Erlenmeyer flasks be-
    fore autoclaving at 15 psi for 15 min.

2. Maintain all cultures at 25°C in 12 hrs of alternating light
    (approx. 5 ft. c.) and 12 hrs of darkness.

3. Dissolve β-sitosterol by one of the methods described on
    page 83.

MERZ, W.G., R.G. BURRELL and M.E. GALLEGLY. 1969. Phytopathology 59: 367-370.

VI. Growth of *P. cinnamomi*

                                                      per liter

Sucrose . . . . 20 g

L-glutamic acid . . . . 2.5 g

KH_2PO_4 . . . . 1.0 g

$Ca(NO_3)_2$ . . . . 0.5 g

KCl . . . . 0.5 g

$MgSO_4 \cdot 7H_2O$ (dissolve separately and add last to minimize
precipitation of calcium phosphate) . . . . 0.5 g

Fe-EDTA . . . . 0.6 ml

(gives 6 ppm Fe^{2+} final concentration)

Micro-elements

    1 ml of a stock solution to give:

$B^+(H_3BO_3)$ . . . . 0.5 ppm

$Mn^{2+}(MnSO_4 \cdot 4H_2O)$. . . . . 0.05 ppm

$Zn^{2+}(ZnSO_4 \cdot 7H_2O)$ . . . . 0.02 ppm

$Cu^{2+}(CuSO_4 \cdot 5H_2O)$. . . . . 0.02 ppm

$Mo^{6+}(NH_4)_6Mo_7O_{24} \cdot 4H_{20})$ . . . . 0.02 ppm

Thiamine . . . . 0.5 mg

Agar . . . . 12 g

Distilled water . . . . 1000 ml

Dissolve cholesterol in ether and add at the rate of 50 mg
cholesterol per liter of medium.

Adjust pH of medium to 6.3-6.5 with 1N NaOH.

CHEE, K.H. and F.J. NEWHOOK. 1965. New Zealand J. Agr. Res. 8: 523-529.

VII. Growth of *P. cryptogea, P. drechsleri, P. boehmeriae, P.
     nicotianae* var. *parasitica, P. megasperma & P. cinnamomi*

    1. Basal medium:

L-asparagine . . . . 2.0 g

$MgSO_4 \cdot 7H_2O$ . . . . 0.1 g

$FeSO_4 \cdot 7H_2O$ . . . . 0.001 g

KH_2PO_4 . . . . 0.43 g

K_2HPO_4 . . . . 0.30 g

Thiamine HCl . . . . 25-1000 μg

1 ml of a minor element mixture which provided, in the final solution, 1 ppm of Zn($ZnSO_4.7H_2O$) and 0.02 ppm of
Cu($CuSO_4.5H_2O$), Mo($NaMoO_4.2H_2O$), and Mn($MnCl_2.4H_2O$).

Sucrose . . . . 10.0-15.0 g

Deionized distilled water . . . . 1000 ml

2. pH adjusted to 6.3-6.5.

ERWIN, D.C. and H. KATZNELSON, 1961. Can. J. Microbiol. 7: 15-25.

For growth of *P. capsici*, *P. citrophthora* and *P. palmivora*, the
above medium was modified by the addition of 0.03 g of β-sito-
sterol and 5.3 g of 2-(n-morpholino)-ethanesulfonic acid (MES)
as the buffer.

MITCHELL, D.J. and G.A. ZENTMYER. 1971. Phytopathology 61: 787-791.

VIII. Synthetic liquid growth medium: *P. fragariae*

1. Sucrose . . . . 15 g

    Casein hydrolysate, enzymatic with 13% total N. . . 3.39 g

    $MgSO_4.7H_2O$ . . . . 0.1 g

    $K_2H PO_4$ . . . . 0.3 g

    KH_2PO_4 . . . . 0.43 g

    $ZnSO_4.H_2O$ . . . . 1 ppm

    $FeSO_4.7H_2O$ . . . . 1 ppm

    $CuSO_4.5H_2O$ . . . . 0.02 ppm

    $CaCl_2.2H_2O$ . . . . 0.01 g

    Distilled water . . . . 1 l

2. Add the microelements as 1 ml of stock solution per liter of
    medium.

MAAS, J.L. 1972. Mycopathol. Mycologia applicata 48: 323-324.

IX. Growth of P. *infestans* in synthetic media

Basal medium:

D-glucose . . . . 5 g

L-asparagine (anhydrous) . . . . 0.5 g

KH_2PO_4 . . . . 1 g

$MgSO_4 \cdot 7H_2O$ . . . . 0.5 g

Zn^{+2} . . . . 0.2 mg

Ca^{+2} . . . . 10 mg

Cu^{+2} . . . . 0.02 mg

Mn^{+2} . . . . 0.05 mg

Mo^{+6} . . . . 0.01 mg

β-sitosterol . . . . 10 mg

Thiamine . . . . 100 μg

De-ionized distilled H_2O . . . . 1000 ml

1.0 ppm Ferric iron

200 mg Ascorbic acid

CUPPETT, VIDA M. and V.G. LILLY. 1973. Mycologia 65: 67-77.

X. Growth of P. *nicotianae* var. *parasitica*

	per liter
Glucose	21.6 g
Asparagine	5.8 g
KH_2PO_4	1.0 g
$MgSO_4 \cdot 7H_2O$	0.5 g
$CaCl_2 \cdot 2H_2O$	100 mg

Microelements:

Fe^{3+} as chloride or as sequestrene (sodium ferric diethylen
triamine penta acetate . . . . 1 mg

Zn^{2+} . . . . 1 mg

Cu^{2+} . . . . 100 mg

Mn^{2+} . . . . 10 μg

Mo^{6+} as $(NH_4)_6Mo_7O_{24}\cdot 4H_2O$ . . . . 10 μg

EDTA (ethylene diamine tetraacetic acid), disodium salt .

. . . . 10 mg

HENDRIX, J.W., S. GUTTMAN, and D.L. WIGHTMAN. 1969. Phytopathology 59: 1620-1624.

XI. Ribeiro's medium for oospore production

                                                    per liter
1. Glucose . . . . 4.5 g

   L-asparagine . . . . 0.1 g

   KNO_3 . . . . 0.15 g

   KH_2PO_4 . . . . 1.0 g

   $MgSO_4\cdot 7H_2O$ . . . . 0.5 g

   $CaCl_2$ . . . . 0.1 g

   β-sitosterol (dissolved in 30 ml CH_2Cl_2) . . . . . . 30 mg

2. Microelements

   $Na_2MoO_4\cdot 2H_2O$ . . . . 41.1 mg

   $ZnSO_4\cdot 7H_2O$ . . . . 87.8 mg

   $CuSO_4\cdot 5H_2O$ . . . . 7.85 mg

   $MnSO_4\cdot H_2O$ . . . . 15.4 mg

   $Na_2B_4O_7$ . . . . 0.5 mg

   Above microelements dissolved in 100 ml deionized distilled water.

. $FeCl_3\cdot 6H_2O$ (50 mg) was dissolved in 100 ml deionized distilled H_2O containing 2.6 g EDTA ((ethylenedinitrilo) tetra-
  acetic acid) and 1.5 g KOH to keep the Fe^{+++} in solution.

. Dispense 1 ml of the above microelements made up as a stock
  solution (from #2 above), into 1 liter of the medium.

. Dispense 1 ml of the $FeCl_3$ from the stock solution (#3 above)
  into medium.

6. Adjust pH to 6.2 with 6N KOH and autoclave, 15 min. at 15 psi.

7. Add Difco special noble agar (14 g).

8. After sterilization, add 1 ml of a sterile stock thiamine solution so that 1 ml = 1.0 mg/liter thiamine. Thiamine is sterilized by Millipore filtration.

9. Shake flask to disperse thiamine into the medium and pour into Petri plates.

Notes:

a) This medium has successfully induced formation of oospores in 20 *Phytophthora* spp.

b) In liquid form, this medium readily induces oospore formation of homothallic species; heterothallic species produce fewer oospores in liquid than in solid medium.

RIBEIRO, O.K., D.C. ERWIN, and G.A. ZENTMYER. 1975. Mycclogia 67: 1012-1019.

XII(a) Production of oospores of *Phytophthora* spp. (Leal's medium):

1. Basal Medium: per liter

    KH_2PO_4 . . . . 0.5 g

    $MgSO_4 \cdot 7H_2O$ . . . . 0.5 mg

    $Fe(NO_3)_3 \cdot 9H_2O$ . . . . 0.1 mg

    $ZnSO_4 \cdot 7H_2O$ . . . . 0.2 mg

    $CaCl_2 \cdot 2H_2O$ . . . . 10 mg

    $CuSO_4 \cdot 5H_2O$ . . . . 0.02 mg

    $MnSO_4 \cdot H_2O$ . . . . 0.05 mg

    Thiamine hydrochloride . . . . 1 mg

    Distilled water . . . . 1000 ml pH adjusted to 6.0 with 6N KOH or H_2SO_4, before adding agar.

Agar . . . . 15 g

The carbon and nitrogen ratios are as follows: For *P. cambivora*, *P. palmivora*, *P. cactorum*, *P. hevae* and *P. megasperma*, use glucose at 5 g/l and asparagine 0.1 g/l (C/N ratio 94:1), β-sitosterol (mixed in ether), 10 mg/l.

For *P. citricola*, *P. megasperma*, *P. cambivora*, *P. cinnamomi*, *P. heveae*, *P. cactorum*, *P. ilicis*, *P. capsici* and *P. palmivora*, substitute asparagine with an equivalent amount of KNO_3.

LEAL, J.A., GALLEGLY, M.E. and V.G. LILLY. 1967. Mycologia 59: 953-964.

XII(b) Production of oospores (Leal *et al's* medium)
     P. cactorum and *P. heveae*

		per liter
1.	Sucrose	30 g
	$Ca(NO_3)_2$	0.05 g
	KH_2PO_4	0.6 g
	$MgSO_4.7H_2O$	0.3 g
	L-asparagine	2.0 g
	L-tyrosine	0.5 g
	Thiamine	1 mg
	Agar	15 g
	Distilled water	1000 ml

To 1 liter is added 1 ml from the following soln. in 100 ml.

$FeSO_4.7H_2O$	50 mg
$ZnSO_4.7H_2O$	50 mg
$CuSO_4.5H_2O$	2 mg
$MnCl_2.4H_2O$	2 mg
$NaMO_4.2H_2O$	2 mg

pH 5.5 before autoclaving.

The sugar and calcium are autoclaved separately from the other constituents.

β-sitosterol . . . . 20 mg

LEAL, J.A., J. FRIEND, and P. HOLLIDAY. 1964. Nature (Lond.) 203: 545-546.

XIII. Growth and oospore production: *P. cactorum* (Elliott's medium)

The basal medium contains:

Sucrose . . . . 10.0 g

L-asparagine . . . . 1.0 g

KH_2PO_4 . . . . 0.5 g

$MgSO_4.7H_2O$ . . . . 0.25 g

Trace element soln. . . . . 1 ml

Thiamine hydrochloride . . . . 1 mg

De-ionized water . . . . 1000 ml

Trace element soln. contains in 1000 ml water:

$Na_2B_4O_7.10H_2O$ . . . . 88 mg

$CuSO_4.5H_2O$ . . . . 393 mg

$Fe_2(SO_4)_3.9H_2O$ . . . . 910 mg

$MnCl_2.4H_2O$ . . . . 72 mg

$Na_2MoO_4.2H_2O$ . . . . 50 mg

$ZnSO_4.7H_2O$ . . . . 4403 mg

Ethylenediaminetetraacetic acid disodium salt, 5 g.

Calcium chloride 0.1 g/l (anhydrous salt).

Cholesterol in ether 200 mg/l.

ELLIOTT, C.G. 1972. Trans. Br. Mycol. Soc. 58: 169-172.

XIV. Production of oospores of *P. palmivora* (Huguenin and
Boccas's medium)

In one liter of deionized water:

K_2HPO_4 . . . . 0.1 g

KH_2PO_4 . . . . 0.8 g

K_2SO_4 . . . . 0.5 g

$MgSO_4.7H_2O$ . . . . 0.5 g

$CaCl_2.2H_2O$ . . . . 0.2 g

KNO_3 . . . . 1.0 g

D-glucose . . . . 10.0 g

Thiamine (chlorhydrate) . . . . 2.0 mg

B-sitosterol . . . . 20 mg

2 ml of minor elements consisting of the following minor
elements per liter of medium:

$FeSO_4.7H_2O$ . . . . 0.5 mg

$ZnSO_4.7H_2O$ . . . . 0.5 mg

$CuSO_4.5H_2O$ . . . . 0.02 mg

$MnCl_2.7H_2O$ . . . . 0.02 mg

$TiSO_4.5H_2O$ . . . . 0.02 mg

$Mo_7(NH_4)6.4H_2O$ . . . . 0.07 mg

De-ionized H_2O . . . . 1000 ml

pH 5.5

Agar-agar 20 g.

HUGUENIN, B. and B. BOCCAS. 1971. Annales Phytopathol. 3: 353-371.

Section 4:

Sporangia and zoospore production

A. The following factors influence sporangia production of
    Phytophthora species

1. Moisture

    The presence of water is critical for sporangia formation.
    Some species will only form sporangia in water, while in
    others, moistening the mycelial mats is sufficient to in-
    duce production of sporangia. Changing the water or frequent
    rinsing enhances sporangia formation. However, Duniway (Can.
    J. Botany 53: 1270-1275, 1975), found that *P. capsici* rarely
    formed sporangia in saturated soils. A more moderate wetting
    of the soil may be better for sporangia formation of some
    species.

2. Temperature

    Each species has a minimum, optimum, and a maximum tempera-
    ture at which sporangia will form. See temperature require-
    ments of some *Phytophthora* spp. listed in the following
    pages.

3. Nutrition

    The presence of sterols is essential for sporangia formation.

4. Oxygen

    Atmospheric oxygen is important; sporangia are most abundant
    at the surface of the culture, or substrate. Mitchell and
    Zentmyer (Phytopathology 61: 807-812, 1971), have shown that
    sporangia formation is reduced by oxygen concentration less
    than that of air, and by CO_2 levels greater than that of
    air. Optimum sporulation occurs in air.

5. Light

    Varies with the species. Most species are stimulated by
    light, particularly in the near-UV-blue region of the spec-
    trum (320-475 nm).

6. Age of Culture

    The age of the culture influences the production of sporan-
    gia; the older the culture, the less likely that sporangia
    will form.

7. pH

    The pH of the medium is critical for sporangia production
    of some species. In general, high pH results from the accum-
    lation of ammonia in the medium which is toxic to sporangia

8. Nature of Medium

    Some media are better for sporulation than others. This of-
    ten varies with individual isolates and species.

9. Cu^{++}

    The presence of Cu^{++} ions (as low as 0.01 mM) is inhibitory
    to sporangia formation.

Caution

A. Make sure that the water used to induce sporangia formation
    does not contain copper or chlorine.

B. Transferring agar cultures to soil leachate or pond water
    (a very common practice) runs the risk of the presence of

other *Phytophthora* or *Pythium* species as contaminants. This
will make positive identification of the *Phytophthora* isolate under consideration difficult.

Useful Reference

WATERHOUSE, G.M. 1931. The production of conidia in the genus *Phytophthora*.
Trans. Brit. Mycol. Soc. 15: 311-321.

Sporangia production

Some methods for sporangia production of several *Phytophthora*
species are given in the following pages. Detailed treatment of
methods to induce the formation of sporangia of several *Phytophthora* species can also be found in J. Tuite's book "Plant Pathol-
ogical Methods," Burgess Publishing Co., Minneapolis, Minnesota, 55415, U.S.A. (1969). See Table 15, page 155, and Table 16,
page 157 in Tuite's book.

In general, most species will produce sporangia if the culture
is transferred to a weak nutrient solution (usually 1/5th of
normal strength), or distilled water, and placed under continuous light conditions. Flooding a petri dish culture with distilled water so that the mycelium is completely submerged, often
induces the formation of sporangia.

Methods for Sporangia Production: *Phytophthora* spp.

I. Grass leaves

    1. Cut leaves of bent grass (*Agrostis alba*) or blue grass
       (*Poa pratensis*), into 1 inch (2-3 cm) lengths and boil
       in distilled water for 10 min.

    2. Cool the grass pieces and place on an agar culture of the
       fungus for 24 hrs.

    3. Transfer the grass pieces to a petri plate with 10-15 ml
       of Schmitthenners's salt solution (page 107).

Table 5: Representative times for sporangial production of some *Phytophthora* spp.

Species	Sporangial Production	Authority
P. *cactorum*	☆3 1/2 days	Waterhouse (1931)
P. *cactorum*	12-48 hr	Buddenhagen (1958)
P. *capsici*	36-48 hr	Katsura et al. (1968)
P. *cinnamomi*	8-36 hr	Chen & Zentmyer (1970)
P. *cinnamomi*	48 hr	Chee & Newhook (1966)
P. *citrophthora*	12-15 hr	Fawcett & Klotz (1934)
P. *colocasiae*	24 hr	Waterhouse (1931)
P. *cryptogea*	5 days	Waterhouse (1931)
P. *fragariae*	24-48 hr	Converse & Scott (1962)
P. *infestans*	7 hr	McAlpine (1910)
P. *infestans*	1 week	Hodgson & Grainger (1963)
P. *meadii*	5-7 days	Peries & Fernando (1966)
P. *melonis*	36-48 hr	Katsura et al. (1968)
P. *megasperma* (some isolates)	12-24 hr	Ribeiro (1977)
P. *megasperma* var. *sojae*	72 hr	Hilty & Schmitthenner (1962)
P. *mexicana*	48 hr	Hotson & Hartge (1923)
P. *palmivora*	36-48 hr	Katsura et al. (1968)
P. *nicotianae* var. *parasitica*	24-48 hr	Waterhouse (1931)
P. *phaseoli*	24 hr	Goth & Wester (1963)

☆Times represent period after transfer of cultures to sporangia initiation media.

4. Sporangia usually develop in 48 hrs. Oospores may develop later.

5. Induce zoospore differentiation and release by chilling the **sporangia-laden** grass leaves for 30 min. in a refrigerator.

II. Hemp seed

Hemp seed (*Cannabis sativa*) can also be used to induce sporangia formation of *Phytophthora* spp. The seed should be boiled for 30 min. and treated as I above.

Hemp seed can be obtained from Carolina Biological Supply
Co., Burlington, N.C. 21275   U.S.A.

Note:

Grass leaf cultures are better for sporangia production of *Pythium* spp., while hemp seed is better for *Phytophthora* spp.

SCHMITTHENNER, A.F. 1973. Proc. First Woody Ornamental Disease Workshop 1:
94-110.

III. Schmitthenner's salt solution for Sporangia formation

   Solution A

   Distilled water . . . . 950 ml

   KH_2PO_4 . . . . 0.075 g

   K_2HPO_4 . . . . 0.075 g

   $MgSO_4.7H_2O$ . . . . 0.05 g

   $ZnSO_4.7H_2O$ . . . . 0.0022 g

   $FeSO_4.7H_2O$ . . . . 0.0005 g

   $MnCl_2.4H_2O$ . . . . 0.000035 g

   Solution B

   $Ca(NO_3)_2.2H_2O$ . . . . 0.1 g

   Distilled water . . . . 50.0 ml

Autoclave solution A and solution B separately and mix before
use.

SCHMITTHENNER, A.F. 1973. Proc. First Woody Ornamental Disease Workshop
94-110.

Production of Sporangia of *Phytophthora* spp.

IV. Petri's Medium

    Calcium nitrate . . . . 0.4 g

    Magnesium sulphate . . . . 0.15 g

    Potassium acid phosphate . . . . 0.15 g

    Potassium chloride . . . . 0.06 g

    Water . . . . 1000 ml

Place small pieces of inoculum into the solution at 12-20°C. Replace with distilled water after 3-4 days.

Comments

Some species produce more sporangia in non-sterilized solution.

TUCKER, C.M. 1931. MO. Agr. Exp. Sta. Bull. 153: 208.

V. Wills Mineral Salts Medium

    Method

    KCl . . . . 50 mg

    KH_2PO_4 . . . . 100 mg

    $MgSO_4.7H_2O$. . . . . 50 mg

    K_2CO_3 . . . . 300 mg

    Distilled water . . . . 1000 ml

    pH alkaline

Transfer inoculum from potato dextrose agar to mineral solution

WILLS, W.H. 1954. J. Elisha Mitchell Soc. 70: 235-243.

VI. Steam a 10 oz. (283 g) package of frozen lima beans in 1
    liter of distilled H_2O for 20-30 min., filter through a
    cotton towel or cheese cloth and adjust volume to 2 liters.
    Add 20 g agar/liter and autoclave 15 min. at 121°C, and pour
    into petri-plates.

  1. Inoculate plates with minced inoculum from cultures grow-
     ing in liquid lima bean medium for 5-6 days in continu-
     ous white light.

  2. Place under fluorescent lights (500-700 ft. c) at 20-
     25°C.

  3. Sporangia are produced within 3-4 days.

Comments

Sporangial production has been reported with this method for *P.
cactorum*, *P. heveae*, *P. capsici*, *P. hibernalis* and *P. syringae*.

HARNISH, W.N. 1965. Mycologia 57: 85-90.

VII. Make up the following medium:

    Blended frozen peas . . . . 160 g

    Sucrose . . . . 5 g

    Agar . . . . 15 g

    Distilled water . . . . 1000 ml

  1. Sterilize above medium and inoculate with *Phytophthora*
     spp.

  2. Incubate cultures for 3 to 7 days.

  3. Incubate discs from above cultures in pond water, filtered
     through Whatman No. 1 filter paper, for 2 days at 20°C.

  4. Zoospore release is stimulated by chilling for 30 min. at
     4°C, and returning to room temperature.

DANCE, M.H., F.J. NEWHOOK, and J.S. COLE. 1975. Plant Dis. Reptr. 59: 523-
527.

C. Methods for Sporangia Production of Specific *Phytophthora* spp.

I. *P. capsici, P. palmivora,* and *P. melonis*

Method I

1. Place 30 ml of V-8 juice medium in 100 ml Erlenmyer flask and sterilize.

2. Inoculate medium with 5 mm diameter disk of mycelium growing on PDA agar for 1 week at 28°C.

3. After 3 weeks incubation at 28°C, remove mycelial mat from flask, rinse with sterilized, deionized water through a Buchner funnel.

4. Place clean mycelial mat on 2 sheets of moistened filter paper in a Petri dish.

5. Cover dish with a paper lid and place in a 30°C incubator illuminated with fluorescent lights.

6. Numerous sporangia form on the surface of the mycelial mat in 36-48 hours.

Comments:

1. A paper lid is important to keep the surface of the mycelial mat in a relatively dry condition.

2. This method is not applicable to *P. parasitica, P. citroph-thora* or *P. heveae.*

KATSURA, K., Y. MIYATA and T. MITANI. 1968. Sci. Rept. 20: 32-36. Kyoto Prefectural Univ. Agr.

II. *P. capsici, P. palmivora*

Method II

1. Place 5 mm inoculum plug on a synthetic medium (Ribeiro's

2. Place petri dishes under black fluorescent lamp (15 watt 30 cm from lamp).

3. Abundant sporangia form in 3 days.

RIBEIRO, O.K. 1975. Department of Plant Pathology, Univ. of California, Riverside.

Method III

For synchronous sporangia formation of *P. capsici* see: Yoshikawa, W. and H. Masago. 1977. Can. J. Botany 55: 840-843.

III. *P. cinnamomi*

a) Chen and Zentmyer's technique

1. Prepare a pea broth, or a V-8 juice broth.

   a) Pea broth - Add 500 ml deionized water to 200 g frozen green peas and blend for 5 min. Centrifuge the mixture for 10 min. at 4080 x g. Decant supernatant and add deionized water to make 1 liter

   b) V-8 juice broth - Thoroughly mix 200 ml V-8 juice with 2 g of calcium carbonate. Centrifuge as for pea broth. Dilute supernatant with deionized water to make 1 liter.

2. For sporangium production, dilute the pea and V-8 juice broths to 1/8 and 1/10 respectively with water.

3. Transfer inoculum pieces (200-400) ca. 1-2 mm^3 to a petri plate containing pea or V-8 juice broth.

4. 16-18 hrs later, wash the young mycelial mats with an autoclaved mineral salt solution composed of the following:

$$Ca\ (NO_3)_2\ \ldots\ldots\ 0.01\underline{M}$$

$$KNO_3\ \ldots\ldots\ldots\ 0.005\underline{M}$$

$$MgSO_4\ldots\ldots\ldots\ 0.004\underline{M}$$

☆Chelated iron.....1 ml

Deionized water.1000ml

☆Chelated iron solution contains:

Ethylene-dinitrotetraacetic acid (EDTA) . . 13.05 g

KOH . . . . 7.5 g

$FeSO_4 \cdot 7H_2O$ . . . . 24.9 g

Deionized water . . . . 1000 ml

5. Wash the mycelial mats with 15-20 ml of solution at least 4 times at 1-hr intervals. Drain solution thoroughly from the dish.

6. Incubate petri plate under two 40-watt fluorescent cold day-light lamps suspended 40 cm above the colonies, at $24^{\circ}C$.

7. Sporulation begins within 8 hrs. from time of first washing and reaches a maximum in 24-36 hrs.

8. For zoospores: Place the sporangia at about $5^{\circ}C$ for 15-20 min. and return to $24^{\circ}C$. Zoospores release in approximately 1 hr after chilling.

9. This has proved to be a reliable method for most isolates of P. *cinnamomi* tested.

CHEN, D.W. and G.A. ZENTMYER. 1970. Mycologia 63: 397-402.

b) Cellophane Technique

1. Cut 10 to 15 discs (6 mm diam.) from a 2 to 3 day-old culture of P. *cinnamomi* growing on 20% V-8 agar (per liter: V-8 juice 200 ml; $CaCO_3$ 2 g; agar 20 g).

2. Place inoculum discs on a sterilized disc of washed, uncoated cellophane (90 mm diam.), laying on V-8 juice agar medium.

3. Incubate for 24 hrs at $24^{\circ}C$.

4. Remove the cellophane membrane containing the mycelia and place in a petri dish containing 25 ml of 5% clarified V-8 juice broth (per liter: V-8 juice 50 ml; $CaCO_3$ 2 g; centrifuge at 1,500 rpm for 5 min.).

5. Abundant young mycelia are produced on the cellophane after 24 hrs.

6. Drain broth from petri dish and rinse mycelia on the cellophane with 20 ml of Chen and Zentmyer's mineral solution (see Method IIa, above), twice, then incubate in 20 ml of the mineral solution at $24^{\circ}C$ under continuous fluorescent light (cool white, 200 ft-c).

7. Sporangia can be observed after 9 hrs at which time drain the mineral solution to prevent premature release of zoospores.

8. Abundant sporangia are produced in 12 hrs in moist conditions.

9. For zoospore release, wash cultures 3 times with sterile distilled water, chill at $16^{\circ}C$ for 30 min. and return to $24^{\circ}C$.

10. Zoospores are released about 1 hr after chilling.

HWANG, S.C., W.H. KO, and M. ARAGAKI. 1975. Mycologia 67: 1233-1234.

c) Synchronous Mycelium Growth Technique

1. Place 20 mycelial discs (5 mm diam.), of *P. cinnamomi* (growing on V-8 agar), in each of 4 standard petri plates.

2. Add 25 ml cleared, sterilized V-8 broth (diluted 1:50 with sterile water) and incubate at $25-27^{\circ}C$ for 24 hrs.

3. Replace medium with Chen and Zentmyer's salt solution (see Method IIa above), and incubate for 30 min.

4. Replace the mineral salt solution 3 more times with fresh solution at 30 min. intervals.

5. Incubate the mycelial discs in the final wash under fluorescent light (200 μW cm^{-2}) at $25-27^{\circ}C$ for 24 hrs.

6. Sporangia form particularly at the periphery of each mycelial disc.

7. Combine mycelial discs from all 4 plates and replace the salt solution with 25 ml sterile distilled water.

8. Release zoospores by incubating plates at 4°C for 30 min. followed by an incubation at 24°C for 2 hrs.

9. Pipette 1 ml of the zoospore suspension in each of 20 small petri plates (50 mm diam.), containing 5 ml V-8 broth (diluted 1:50), and incubate plates for 30 hrs. A thin mycelial mat of juvenile hypha forms in each plate.

10. Decant the nutrient solution and add 7 ml of the salt solution.

11. After 10 min. replace with 7 ml fresh salt solution and incubate under light for 24 hrs.

12. Sporangia form uniformly in each plate (approx. 150-200 sporangia per 40 x microscopic field), at the end of the 24 hrs.

ZAKI, A.I. and G.A. ZENTMYER. 1975. Phytophthora Newsletter 4: 31-32.

d) Non-sterile Soil Extract

1. Place pieces of fungal culture growing on PDA (made from fresh potatoes) or on V-8 juice agar, in petri plates containing a non-sterile soil extract.

2. Sporangia are produced in 48 hrs at 24°C.

3. Sterilizing soil prevents sporangial formation.

Note:

No sporangia are produced in mycelium growing on cornmeal agar.

ZENTMYER, G.A. and LEE ANN MARSHALL. 1959. Phytopathology 49: 556.

e) Bacterial Stimulation

Adding the bacterium *Chromobacterium violaceum* and several other bacteria and combinations of bacteria to sterile

soil extracts also stimulates sporangial production of *P.
cinnamomi*.

ZENTMYER, G.A. 1965. Science 150: 1178-1179.

Note:

An elaborate method for the induction of aseptic sporangial
formation in *P. cinnamomi* by metabolic diffusates of soil microorganisms has been described by Marx and Haasis, Nature   206:
673-674, (1965).

f) Hemp Seed Technique

   1. Place a few hemp seeds in 20 ml distilled water in a
      250 ml flask. Alternately use 20 ml of the water extract
      of hemp seeds soaked for 24 hrs.

   2. Strip mycelium of *P. cinnamomi* from PDA plates and add
      to flasks.

   3. Incubate at 27°C for 5 days.

Note:

No sporangia are formed in autoclaved preparations under any
light conditions.

MANNING, W.J. and D.F. CROSSAN. 1966. Phytopathology 56: 235-237.

g) Zoospore Formation

   For zoospore formation, take mycelial mats containing
   sporangia and rinse in distilled water. Then place the mats
   in fresh distilled water in petri dishes at 8°C for 15 min.
   Return to room temperature for zoospore release.

IV(a) *P. citrophthora*

1. Grow a culture of the fungus for 10 days at $20^{\circ}C$ on 2%
   potato dextrose agar (PDA), or 2% oatmeal agar medium.

2. Cover with tap water or with 0.01M KNO_3 and incubate at
   $25^{\circ}C$.

3. After 4 to 5 days, sporangia begin to develop in the
   cultures and within 7 to 8 days, abundant sporangia are
   obtained.

IV(b)

1. Grow a culture of the fungus for 4-6 days at $25^{\circ}C$ in a
   liquid medium of 20% PDA, 12% dextrose, or 5% oatmeal.

2. Wash mycelial fragments from this culture in tap water
   and transfer to a petri dish containing tap water or
   0.01M KNO_3.

3. Incubate at $28^{\circ}C$.

4. Sporangia begin to appear on the fragments of mycelium
   after 1 to 2 days and are abundant within 3 days.

Note:

Sporangia by these two methods can also occur at sub-optimal
temperatures, but the process is prolonged.

SCHIFFMAN-NADEL, M. and E. COHEN. 1968. Phytopathology 58: 550.

V. *P. citrophthora, P. nicotianae* var. *parasitica* and *P. cac-
    torum*

   Method I

   1. Grow fungus in a weak prune-juice broth (60 g prune pulp
      in 1 liter tap water) at $2-4^{\circ}C$ below its optimum temp.
      for 10 days - 2 weeks.

2. Transfer mycelial mat to petri dish containing sterile tap water, so that approx. half the mycelial mat is submerged.

3. Incubate in light. Sporangia begin to form in 12-15 hrs for *P. citrophthora*.

4. For zoospore release, float petri dish for 3-5 min. in water bath at 28-30°C.

5. Transfer petri dish to water bath at 15-18°C. for 1 hr.

FAWCETT, H.S. and L.J. KLOTZ. 1934. Phytopathology 24: 693-694.

Method II

1. Place 3" lengths of alfalfa (Lucerne) stems in testtubes containing 0.5 g $CaCO_3$ and deionized water to a depth of 2", and autoclave.

2. Place bits of mycelium on the stems at the water surface.

3. Incubate at 26°C for 4 to 5 days, resulting in abundant mycelium.

4. Remove stems and place on a piece of cloth wrapped around the top of a container filled with water.

5. Allow a steady stream of water to fall from a height of 1-2 ft. onto the stems, to aerate the water and the cultures.

6. In 4-5 hrs at a water temp. of 20-25°C, *P. citrophthora* begins to form sporangia. Sporangia are abundant in 48 hrs.

7. At water temp. of 25-27°C, *P. parasitica* forms sporangia.

8. Placing the sporangia-laden alfalfa stems at 23-24°C results in zoospore release of *P. parasitica*, 16-18°C *P. citrophthora*.

KLOTZ, L.J. and T.A. DeWOLFE. 1960. Pl. Dis. Reptr. 44: 572-573.

VI. Sporangia production of *P. drechsleri*

1. Surface sterilize one month-old seedlings of *Eucalyptus sieberi* for a few seconds in 70% ethanol.

2. Float the sterilized seedlings in petri plates containing pieces of mycelium from cultures of *P. drechsleri* incubated in darkness at 25 $\pm$ 1°C for 7 days.

3. Incubate the petri plates under lab light conditions at 23-28°C.

4. Abundant sporangia form in 2-5 days on the cotyledons, hypocotyls, roots, and on the agar close to the seedlings when agar cultures are used.

5. This technique can also be used for sporangia production of *P. cactorum* and *P. cryptogea*.

6. This technique is not effective for sporangia production of *P. megasperma* and *P. cinnamomi*.

GERRETTSON-CORNELL, L. 1975. Aust. Plant. Pathol. Soc. Newsletter 4(3): 32.

VII. *P. erythroseptica*

1. Grow fungus on bean agar for 5-7 days.

2. Remove and incubate for 2-3 days at $\pm$ 18°C in distilled water, pond water or compost percolate.

BYWATER, J. and C.J. HICKMAN. 1959. Br. Mycol. Soc. Trans. 42: 513-524.

VIII. *P. fragariae*

Method I

1. Place 2-3 week-old cultures growing on lima bean in non-sterile tap water.

2. After 24-48 hrs at 13°C, abundant sporangia develop.

CONVERSE, R.H. and D.H. SCOTT. 1962. Phytopathology 52: 802-807.

Method II

1. Autoclave seeds of sunflower, squash, pea, or soybean in tap water.

2. Inoculate seeds with 2 to 3 week-old fungus cultures growing on oatmeal-extract agar.

3. Incubate for 2-3 weeks at 20°C.

4. Transfer the fungus-covered seeds to petri plates, containing shallow tap water and rinse daily.

5. Abundant sporangia are produced in 4-5 days.

ELIX, E.L. 1962. Phytopathology 52: 9.

X. *P. infestans*

Method I

Steam rye seed (100 g) in water for 30 min. and filter through cheese-cloth. Add 5 g dextrose and dilute extract with equal volume of water. Add 1.5% agar, bring volume up to one liter with distilled water, and autoclave.

ote:

porulation is obtained in approx. 1 week at 20°C.

ODGSON, W.A. and P.N. GIRAINGER. 1963. Can. J. Plant Sci. 44: 583.

Method II

1. Make up lima bean agar (see page 76) and pour into 25 ml test-tubes.

2. Autoclave and slant.

3. Inoculate with fungus and incubate at 18°C.

4. Abundant sporangia form within 8 days.

Notes:

Sporangial production obtained with all races of *P. infestans*
tested. For zoospore formation and release, float sporangia in
distilled water in petri plates at 8-10°C. Zoospores are released within 1 1/2 - 2 hrs.

GALLEGLY, M.E. W. Va. Univ. Morgantown.

X. *P. meadii*

1. Prepare lima bean agar medium and adjust pH to 6.

2. Inoculate with *P. meadii*.

3. Incubate at 28-30°C for 5-7 days.

4. Zoospores are readily released in free water between
   20-28°C.

PERIES, O.S. and T.M. FERNANDO. 1966. Trans. Brit. Mycol. Soc. 49: 311-325.

XI. *P. megasperma* isolates

Studies on numerous isolates of *P. megasperma* from several different hosts indicate that no single technique is effective for
sporangia production of all isolates tested in axenic culture.

The three techniques described below have, among them, success
fully induced sporangia formation of most *P. megasperma* isolates tested.

Method I

1. Cut 5 mm diameter discs from the edge of a 5-day-old
   fungus colony growing on V-8 juice agar.

2. Transfer discs to petri dishes containing 15 ml each of
   diluted V-8 juice (1 part V-8: 4 parts sterile distille
   H_2O).

3. Incubate for 24 hrs. at 25°C under continuous fluoresce
   light (250 ft C).

4. Decant V-8 juice and replace with 15 ml sterile distilled water.

5. Incubate under lights for a further 24 hrs at 25°C. Numerous sporangia and swimming zoospores can now be observed in the petri dishes.

Note:

The water can be replaced in the dishes and incubated under lights overnight if abundant sporangia formation is desired.

Method II

1. Grow cultures in petri dishes containing 10 ml each of either clarified V-8 juice solidified with agar or grow on Ribeiro's synthetic agar medium until colony reaches the edge of the plate.

2. Flood the plates with sterile distilled water and place on a shaker at 60 excursions/min. for 48 hrs.

3. Remove plates, change the water, and place at 12°C overnight.

4. Sporangia and swimming zoospores can now be observed in the dishes.

Method III

1. Grow cultures in petri dishes containing 15 ml each of either liquid V-8 juice or Ribeiro's synthetic liquid medium until colonies reach the edge of the plates.

2. Decant the medium and rinse the fungal colonies twice with sterile distilled water.

3. Incubate in sterile distilled water at 12°C or at room temperature overnight.

4. Sporangia and swimming zoospores can then be observed in the petri dishes.

RIBEIRO, O.K. and J.S. BAUMER. 1977. Phytophthora Newsletter 5: 42-43.

Method IV

1. Grow culture on a V-8 juice medium (adjusted to pH 6 with $CaCO_3$), for 10-20 days.

2. Cut mycelial discs from the agar medium and place in an inorganic nutrient soln. (pH 5) consisting of: KH_2PO_4, 0.001 M; KNO_3, 0.005 M; $Ca(NO_3)_2.4H_2O$, 0.005 M; $MgSO_4.7H_2O$, 0.002 M; microelements: 1 ml/liter of the following stock soln. H_3BO_4, 2.86 g; $MnCl_2.7H_2O$, 0.22 g; $(NH_4)_6 Mo_7O_{24}.4H_2O$, 0.2 g in 1 liter of distilled water; 1 ml of a stock soln. of ethylenediamine tetra-acetic acid (EDTA) 26 mg (8.9×10^{-5} M), KOH 1.5 g, $FeSO_4.7H_2O$, 0.24 g in 1 liter of distilled water.

3. Incubate on shaker for 24 hrs at room temp.

4. Copper in the inorganic salt soln. at concentrations as low as 0.2 ppm, Cu^{++} inhibits sporangial production.

KENNEDY, B.W. and D.C. ERWIN. 1961. Trans. Brit. Mycol. Soc. 44: 291-297.

XII. *P. megasperma* var. *sojae*

Method I

1. Grow fungus on Difco cornmeal agar for 10 days at $20^{o}C$.

2. Take 1 x 1 cm squares cut out of the agar, and place in petri's soln. (see p. 108).

3. Wash in distilled water. Sporangia develop shortly there after.

HILDEBRAND, A.A. 1959. Can. J. Botany 37: 927-957.

Method II

1. Autoclave 1 g of dried lima beans in 1 liter distilled water for 30 min. and decant into petri plates.

2. Inoculate with pieces of the fungus and incubate at $25^{o}C$ for 72 hrs.

3. Expose to $14-15^{o}C$ for 8-10 hrs for zoospore production.

HILTY, J.W. and A.F. SCHMITTHENNER. 1962. Phytopathology 52: 859-862.

Method III

1. Place diseased soybean tissues in a sieve fitted over a
   600 ml beaker.

2. Add one drop of Tween 20 and run tap water over the tissues for 20 min. to 2 hrs. Tween 20 (a wetting agent)
   loosens surface materials on the diseased tissue.

3. Place the washed diseased tissues on 2.3 g/l liter of
   Difco lima bean agar in petri plates kept at room temp.

4. Abundant sporangia develop and zoospores are released within 24-48 hrs even in the presence of contamination by
   other fungi and bacteria.

KLEIN, H.H. 1959. Phytopathology 49: 380-383.

XIII. *P. nicotianae* var. *parasitica* (*P. parasitica*)

Method I

Boil 300 g of air-dried soil for 1 hr. Remove particulate
matter by vacuum filtration. Use solution undiluted after
autoclaving in 20-30 ml amounts at 15 psi for 15 min. Use
inoculum plugs from PDA cultures, 3-5 days old.

WILLS, W.R. 1954. J. Elisha Mitchell Soc. 70: 235-243.

Method II

1. Strip 6-20 day-old cultures from oatmeal agar plates,
   and place in dry petri plates.

2. Wet cultures with 0.01 M KNO_3 solution at pH 6.0. Do *not*
   flood the cultures.

3. Incubate 8-12 days at $26^{\circ}C$ in continuous light (250 ft-c).

4. For zoospores, add 10 ml of distilled water to mycelial
   mats bearing abundant sporangia and chill 25 min. at $8^{\circ}C$.

5. Return to room temp. and approx. 15 min. later most of
   the sporangia discharge zoospores.

GOODING, G.V. and G.B. LUCAS. 1959. Phytopathology 49: 277-281.

Method III

1. Make up the following medium:

   V-8 juice . . . . 200 ml

   CaCO$_3$ . . . . 2 g

   Distilled water . . . . 800 ml

2. Strain the mixture first through a metal strainer, then twice through several layers of cheesecloth, and finally through a layer of absorbent cotton. About 700-800 ml should be recovered at the end of this procedure.

3. Filter the medium repeatedly through Whatman #1 filter paper in a Buchner funnel, until clear.

4. Dilute the clear V-8-CaCO$_3$ broth with an equal volume of distilled water.

5. Pipette 15 ml of the medium into a 6-oz prescription bottle and autoclave for 15 min. at 121°C.

6. The pH of medium after autoclaving should be between 6.4 and 6.7.

7. Inoculate the medium in each bottle with a 7 mm mycelial disc taken from the edge of a colony of *P. nicotianae* var. *parasitica* growing on V-8 juice agar (20 g agar/ liter of clear V-8-CaCO$_3$ broth).

8. Incubate the bottles for 24 hrs in a 25°C incubator and then shake vigorously by hand (20-30 strokes) to frag- ment the mycelium.

9. Incubate the bottles in a horizontal position for a further 5 days at 25°C. The mycelium should now cover most of the medium surface.

Induction of Sporangia Production

1. Pour the contents of each bottle aseptically in a sterile 35 ml screwcapped centrifuge tube and centrifuge in a clini cal centrifuge for 10 min. at 1610 xg (3200 rpm).

2. Add 20 ml of sterile glass-distilled water (triple distilled water), to the mycelial mat in the centrifuge tube and repeat the process.

3. After the third centrifugation, resuspend the mycelial mat in 10 ml distilled water and pour the entire contents into a deep, flat-bottomed glass petri plate (90 x 20 mm).

4. With the aid of a transfer needle, aseptically remove the original inoculum disc, and spread the mycelial mat out.

5. Incubate the plates for 5 days at 25°C for sporangium formation.

6. Avoid disturbing the cultures before the fifth day to prevent premature swarming of zoospores.

Induction of Zoospore Production

1. Wash the sporangium-bearing mycelium three times by decanting the water from the mycelial mat in the plate, and replacing it with 10 ml portions of distilled water.

2. Incubate the plates at 20°C for 15 min. or longer to induce zoospore swarming, then bring back to room temp. (25 ± 2°C).

3. Thirty minutes after swarming begins, pool the zoospores by transfering to another vessel, using a pipette to effect the transfer.

Note:

Avoid a lengthy swarming period which results in zoospores of varying ages. Such zoospores are undesirable for spore germination studies.

Approximately 200,000-500,000 zoospores/ml are obtained by this method.

A higher concentration of V-8 juice enhances mycelial growth but reduces sporangia production.

This method is also applicable to P. palmivora.

MENYONGA, J.M. and P.H. TSAO. 1966. Phytopathology 56: 359-360.

XIV. *P. palmivora*

Method I

1. Grow cultures at 31°C in a V-8 juice medium consisting of: 10% V-8 juice, 0.2% $CaCO_3$ and 1.5% agar.
2. After 2 days, expose cultures to fluorescent lamp radiation (200 ft.-c) for 24 hrs.
3. *P. cactorum* and *P. colocasiae* produce equal numbers of sporangia under irradiation or in the dark.
4. *P. citrophthora*, *P. cryptogea* and *P. megasperma* produce no sporangia under irradiation or in the dark under the above conditions.

ARAGAKI, M. and R.B. HINE. 1963. Phytopathology 53: 854-856.

Method II

1. Grow cultures for 2 days in the dark at 28°C on vegetabl juice agar (10% V-8 juice, 0.2% $CaCO_3$, 1.5% agar) at 25 ml/petri dish.
2. Induce sporulation by placing the cultures under fluores cent illumination (ca. 200 ft.-c) at 24°C for 2 days.
3. Zoospores are released by placing sporangia in distilled H_2O.

Note:

Approx. 80% or more of the sporangia produce zoospores in the range 16-31°C, with the optimum at 24-48°C.

ARAGAKI, M., R.D. MOBLEY, and R.B. HINE. 1967. Mycologia 59: 93-102.

XV. *P. phaseoli*

1. Place 2 g of Golden Bantam sweet corn seed in a flask
   containing 100 ml of deionized distilled water.

2. Inoculate with cultures growing on lima bean agar medium.

3. Incubate 14 days at 20°C.

4. Remove mycelium covered seeds and expose to an atmosphere
   of 90-100% relative humidity at $18-22^{\circ}$C. Sporangia develop over the entire seed within 24 hrs.

GOTH, R.W. and R.E. WESTER. 1963. Phytopathology 53: 233-234.

XVI. *P. primulae*

1. Grow fungus on bean meal agar. (50 g bean meal, 20 g
   agar, 1 liter distilled water.)

2. Place agar discs in sterile, filtered soil-extract solution at $15-20^{\circ}$C. (100 g soil, 200 ml distilled water,
   shaken and filtered, and the filtrate cleared by centrifugation.)

3. Pass a continuous stream of air bubbles through the solution.

TOMLINSON, J.A. 1952. Br. Mycol. Soc. Trans. 35: 221-235.

. Zoospore Induction and Release

Germination of sporangia can occur in either of two ways:

. Direct i.e. by means of a germ tube(s) that continue(s) to
  elongate and branch to form mycelium. This mode of germination occurs when the culture environment is rich in nutrients, lacking in free water and subjected to ambient or
  higher temperatures.

. Indirect i.e. by cleavage of the protoplasm to form motile
  zoospores which are discharged by the dissolution of the

apical papillum. Indirect germination requires the presence
of free water and a lowering of the temperature in the vici
ity of the sporangium for a majority of the species in the
genus. Usually 15-18°C for 10-20 minutes is sufficient to
induce zoospore differentiation. Returning cultures to room
temperature results in release of zoospores. The zoospores
can then be filtered through Whatman #1 paper to remove the
sporangial shells.

Note:

Zoospore formation is strongly inhibited in some species by
glucose and other catabolites such as citric acid cycle acids,
and amino acids. This inhibition can be partially reversed by
cylic AMP at concentrations of 10^{-6} to 10^{-5} M. Reversal of inhibition by cyclic AMP however, is not possible when caused by
glucose concentrations of greater than 5 mM. Repression of
sporangia production caused by non-catabolites such as antimyc:
A or cycloheximide are not reversed by cyclic AMP.

YOSHIKAWA, M. and H. MASAGO. 1977. Can. J. Botany 55: 840-843.

E.   Zoospore characteristics

I.   Morphology and physiology

   1. Reniform to pyriform in shape, with anterior end tapere

   2. Possess a length-wise groove (see scanning electron mi-
      crographs in Desjardins *et al.* Archiv. Microbiol. 88:
      61-70, 1973).

   3. Zoospores possess two flagella; a tinsel (anteriorly di
      rected) flagellum, and a whiplash (posteriorly directed
      flagellum. The tinsel flagellum has prominent mastigone
      mes; the whiplash flagellum possesses fine lateral and
      tip hairs. Both flagella are borne on the concave side
      of the zoospore in the deep part of the groove.

   4. Zoospores contain golgi apparatus, mitochondria, rough

endoplasmic reticulum, microtubules, microbodies, mem-
brane-free ribosomes, and a large quantity of vesicles
(Hemmes, D.E. and H. Hohl, J. Cell. Sci. 9: 175-191,
1971).

5. Motile zoospores exhibit strong attraction to certain
   plant exudates, particularly certain amino-acids (chemotaxis), and to certain levels of electrical current
   (Electrotaxis). See pages 296-298 for details.

6. Zoospores possess an adhesion layer during the initial
   stage of encystment, for attachment to solid surfaces.
   See below for details.

7. Motile and germinating zoospores are able to catabolise
   glucose, acetate, serine, glutamate and asparagine
   (Barash *et al.*, Phytopathology 55: 1257-1261, 1965).

I. Mobility

Zoospores swim in a helical path with an amplitude of 26-70 μm.
See Fig. 2. Collisions with solid surfaces cause a disorienting
effect and markedly restrict their active movement. Unobstructed locomotion requires pores of approximately 50-140 μm in
diameter (Allen, R.N. and F.J. Newhook, Trans. Brit. Mycol. Soc.
63: 383-385, 1974). Thus, pore size of soil types influence the
mobility of zoospores. Pfender *et al.* (Phytopathology 67: 657-
63, 1977), have shown that zoospores can migrate upward through
15 mm of a sandy loam soil (av. pore size ⟩ 190 μm), and only
4 mm in a silt loam soil (av. pore size ⟨ 120 μm). *P. cinnamomi*
zoospores have been shown to travel 76 mm in soil leachate and
4 mm in wet soil (Kuhlman, E.G., For. Sci. 10: 151-158, 1974).

Ho and Hickman (Can. J. Botany 45: 1983-1994, 1967), have measured the velocity of zoospores of *P. megasperma* var. *sojae* at
56.5 μ/sec at 20°C. Unobstructed, the zoospore can thus travel
63 mm in an hour.

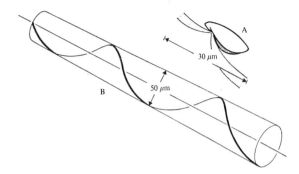

Fig. 2: Diagrammatic representation of a zoospore of *P. cinnamomi* showing
(A) the effective diameter as defined by the distance between the tips of
the fore and aft flagella and (B), the typical helical path of movement
when a zoospore swims in water.

From ALLEN & NEWHOOK (Trans. Brit. Mycol. Soc., Vol. 63, 1974). By permission of Cambridge University Press.

III. Zoospore longevity in vitro

The duration of motility of zoospores depends on the following
factors:

1. Temperature: In general, increasing temperatures result
   in decreased motility of zoospores. The temperature for
   optimum motility varies with the species. Length of
   period of zoospore motility of some *Phytophthora* spp.
   are as follows:

   P. citrophthora - 24 hrs at 12.5^{o}C, 1 hr. at 24°C, and
   a few seconds at 7°C (Klotz, L.J. Phytopathology 30: 14
   1940).

   P. fragariae - 15 min. at 5-6°C, 5 hrs at 13-14°C, 1 1
   hrs at 20°C, and 15-30 min. above 20°C (Goode, P.M.
   Trans. Brit. Mycol. Soc. 39: 367-377, 1956).

   P. infestans - 22 hrs at 3°C, and 30 min. at 24°C
   (Crosier, W. Phytopathology 23: 713-720, 1933).

   P. megasperma var. *sojae* - 24-48 hrs at 24-25°C (Ribei
   unpublished).

P. palmivora (some strains) - maximum motility, 1 1/2 hrs. (average 22 min.) at 17°C, decreases rapidly in water at 20-25°C.

P. phaseoli - 24 hrs at 10°C, 12 hrs at 15°C, 6 hrs at 20°C, and ⟨ 3 hrs at 25°C (Hyre, R.A. and R.S. Cox Phytopathology 43: 419-425, 1953).

2. Presence of water: Zoospores are motile for longer periods of time if swimming in free water.

3. pH: The lower the pH, the lower the motility e.g. for *P. fragariae*, motility is 5 min. at pH 4.4 and 4 hrs at pH 6.8 (Goode, 1956).

4. Mechanical disturbances: *Phytophthora* zoospores rapidly cease swimming and encyst if subjected to vigorous shaking, centrifugation, etc.

5. Contact with solid surfaces: Zoospores quickly decrease their motility on contacting a solid surface e.g. petri plate edge, leaf surface, etc.

6. Concentration of zoospores: A direct correlation has been reported between zoospore concentration and motility. At concentrations of 10^6 zoospores/ml, more than 50% of *P. nicotianae* var. *parasitica* zoospores were motile after 4 hrs. At 10^3 zoospores/ml, motility ceased after 5 minutes (Gooding, G.V. and G.B. Lucas, Phytopathology 49: 277-281, 1959).

V. Adhesiveness of Zoospores

Adhesiveness to a solid surface is a property of the zoospore which only becomes apparent during the initial stage of encystment. The adhesive phase is of very short duration. The adhesive material, which appears to be a glycoprotein, is believed to be discharged by the peripheral vesicles just prior to encystment. This phenomenon was first described by Sing and Bartnicki-Garcia, . Cell Sci. 18: 123-132 (1975).

F. Measuring the concentration of zoospores

The most common method of measuring zoospore concentration is
with a Haemocytometer. Other methods used successfully are the
Levy-Hauser counting chamber, Coulter counter, and the microsyringe method. The first three methods carry their own extensive instructions and are relatively simple to use. The microsyringe method is described below.

Measuring the concentration of Zoospores: Microsyringe techniqu

  1. Pre-adjust to 2-μl the volume control sleeve attached to
     the plunger guide of a 10-μl microsyringe.

  2. Take up a 2-μl aliquot of diluted spore suspension from
     the sample and spread in two 1-μl drops on a glass slide
     and count the number of zoospores in each drop.

  3. The total area of one drop of spore suspension is approx
     mately equal to two microscopic fields at 10X.

  4. The number of zoospores per ml is calculated from the
     average number of zoospores per μliter.

  5. For easy counting, dilute spore suspension to 5 motile
     zoospores/μl.

Notes:

a) Method comparable to haemocytometer.

b) Sporangia and other fungal propagules can be counted.

c) Advantage of this method is that motile zoospores can be
    counted.

KO, W.H., L.L. CHASE, and R.K. KUNIMOTO. 1973. Phytopathology 63: 1206-1207

G. Test for viability of sporangia and zoospores

Method

1. Rose Bengal (300 ppm) can be used to check viability of
   sporangia and zoospores.

2. Dead plasma is indicated by a red coloration; live cells
   remain colorless.

GISI, U. and F.J. SCHWINN. 1976. Microscopa Acta. Band 77. Heft 5: 402-419.

Section 5:

Chlamydospores

A.   Factors affecting production of chlamydospores

   1. Temperature

     Optimum temperature is usually around 15°C, although Chee
     (1973) reported an optimum around 32°C for chlamydospore
     production of *P. palmivora*.

   2. Carbon-nitrogen ratio

     A carbon-nitrogen ratio of 30:1 is optimum for production
     of chlamydospores of *P. palmivora*. Casein hydrolysate was
     found to be the best nitrogen source   (Hueguenin, 1974).
     Ammonium sulfate inhibited formation of chlamydospores
     of *P. palmivora*.

   3. pH

     Conflicting reports exist on the importance of pH for
     chlamydospore formation. Huguenin (1974) found no signi-
     ficant difference between pH 4.5 and 7.0 for chlamydo-
     spore formation, while Chee (1973) reported that a pH of
     5.8 is optimum. Both these investigators used *P. palmi-*
     vora in their studies.

   4. Sterols

     A definite requirement for sterols for chlamydospore for-
     mation has been demonstrated.

   5. Depth of liquid medium

     The depth of the medium influences the quantity of
     chlamydospores formed. Chee (1973) obtained chlamydo-
     spores in shallow water at 15°C, and none in submerged
     mats, while Tsao (1971) and others report that chlamydo-
     spore formation was best in submerged cultures in a deep
     layer of water at 15-18°C.

References:

CHEE, K.H. 1973. Trans. Brit. Mycol. Soc. 61: 21-26.

HUGUENIN, B. 1974. Annales Phytopathol. Paris 6: 425-440.

TSAO, P.H. 1971. Phytopathology 61: 1412-1413.

Notes:

1. The definition, origin and morphological characteristics of
    the different types of chlamydospores of *Phytophthora* are
    given by E. Blackwell, Mycol. Paper No. 30: 23 p. (1949).

2. A liquid synthetic medium e.g. Ribeiro's synthetic medium
    (page 95), with a high carbon-nitrogen ratio will induce
    chlamydospore formation of a number of species.

B. Factors affecting the dormancy and germination of chlamydo-
    spores

    1. Dormancy of chlamydospores is exogenous i.e., germination
       is enhanced by the addition of nutrients (Mircetich *et
       al.*, 1968). In soil, plant root exudates stimulate ger-
       mination of *P. cinnamomi* chlamydospores (Zentmyer *et al.*,
       1973). However, Chee (1973) found that no exogenous nu-
       trients were required for germination of *P. palmivora*
       chlamydospores. Mircetich and Zentmyer (1969) on the
       other hand, reported that germination of *P. cinnamomi*
       chlamydospores was enhanced in glucose or asparagine
       amended soils.

    2. Germination of chlamydospores is best on water agar.
       Cother and Griffin (1974) reported enhanced germination
       of *P. drechsleri* chlamydospores on water agar amended
       with a salt mixture giving an osmotic potential of -3.8
       bars. Germination was obtained as low as -56.0 bars, in-
       dicating that chlamydospore germination can occur under
       relatively dry conditions.

C. Factors found to influence germination of *P. nicotianae*
    var. *parasitica* are as follows:

1. Distilled water inhibits germination.

2. 2% natural soil extracts result in 38-49% germination.

3. Moist, non-amended natural soils (both sterile and un-
   sterile) results in 43-78% germination.

4. Germination is reduced or inhibited when natural or
   sterile soils are amended with L-asparagine or NH_4NO_3.

5. Germination is increased with vegetable juice or excised
   citrus roots.

6. Germination is increased by the addition of individual
   amino-acids e.g., asparagine, alanine or glutamine at
   0.01 M.

7. NH_3NO_4 or sugar (glucose, fructose and sucrose) does
   not increase germination.

These tests were all done on 3-4 week old chlamydospores.

References:

CHEE, K.H. 1973. Trans. Brit. Mycol. Soc. 61: 21-26.

COTHER, E.J. and D.M. GRIFFIN. 1974. Trans. Brit. Mycol. Soc. 63: 273-279.

MIRCETICH, S.M. and G.A. ZENTMYER. 1969. Phytopathology 59: 1732-1735.

MIRCETICH *et al.* 1968. Phytopathology 58: 666-671.

SAO, P.H. and J.L. BRICKER. 1968. Phytopathology 58: 1070.

. Determination of chlamydospore viability

1. Place one drop of a stock spore suspension in the cavity
   of a culture slide, containing 0.3 ml of aqueous rose
   bengal solution at a concentration of 50 ppm.

2. Incubate slides in darkness at 24°C for 6-12 hrs.

3. Injured and non-viable chlamydospores are selectively stained.

4. The percentage of non-stained viable chlamydospores are calculated from the total number of spores examined in each of three replications.

Note:

Selective staining of chlamydospores by rose bengal does not occur in media containing citric acid at concentrations of 0.025 M, or V-8 juice at 100 ml/liter and streptomycin at 50 ppm.

MIRCETICH, S.M., G.A. ZENTMYER, and J.B. KENDRICK Jr. 1968. Phytopahtology 58: 666-671.

E. Detection of colonies arising from germinated chlamydospore

The following antibiotic medium has been used successfully to detect colonies arising from germinated chlamydospores:

    Agar 4

    V-8 juice (see page 75) 1

    Nystatin (mycostatin) . . . . 50 pp

    Vancomycin . . . . 100 pp

    PCNB (pentachloronitrobenzene) . . . . 10 pp

McCAIN, A.H., O.H. HOLTZMAN, and E.E. TRUJILLO. 1967. Phytopathology 57: 1134-1135.

F(a) Concentration of chlamydospores from soil: *P. cinnamomi*

    1. Screen soil through 1/4" (6.35 mm) sieve to remove rock and debris.

2. Store soil above 15% moisture level at $20^{o}C$ to prevent loss of germinability of chlamydospores.

C. Separation of chlamydospores from soil

1. Place 50 g of screened soil in 500 ml of water in a Waring blender.

2. Stir for 1 min. at slow blender speed (approx. 3000 rpm).

3. Immediately pour supernatant through nested sieves with mesh sizes of 149, 61, 44, and 38 μm.

4. Collect supernatant passing through 38 μm sieve.

5. Restir (twice), the sediment left in the blender with the addition of 200 ml water at slow blender speeds for 20 secs.

6. Pour through nested sieves.

7. Wash the residue on the 149 μm sieve.

8. Collect the residue in the 149 μm sieve and in each of the remaining sieves into separate beakers by holding the screen at an 80° angle and carefully backwashing on the bottom side of the screen with a moderate stream of water from a small rubber tubing.

9. Rescreen the suspension which passed through the 38 μm sieve three times through the 38 μm sieve. Hold the sieve at a 60°-70° angle.

10. Collect residue as before.

11. Place sieved chlamydospores on a selective medium (see McCain's technique on page 140), and count the number of colonies formed 24 hrs. later.

Notes:

Sieved chlamydospores germinate readily and grow faster than chlamydospores from unsieved natural soil, on the selective medium.

b) *Pythium* spp. arising from oospores which measure less than
   38 μm are the main source of contaminating colonies. Very
   few *Pythium* colonies however, grow on the selective medium
   when flooded with the concentrated chlamydospore suspension

Advantages:

a) A greater number of chlamydospores are detected per gram of
    soil.

b) Few or no colonies of other fungi develop on the plates.

c) Since the number of bacteria and small-spored fungi are re-
    duced by sieving, lower amounts of antibiotics can be used.

McCAIN, A., O.V. HOLTZMANN, and E.E. TRUJILLO. 1967. Phytopathology 57:
1134-1135.

F(b) Axenic production of chlamydospores: *P. cinnamomi*

1. Prepare V-8 juice broth (see page 75), pH 4.5-4.7, and
   sterilize.

2. Pour 25 ml of the medium into each sterile 90 mm petri
   dish.

3. Pipette 1 ml of a cholesterol/ether solution (50 mg
   cholesterol in 100 ml diethyl ether), on to the surface
   of the medium in each dish, to give a concentration of
   20 ppm of cholesterol.

4. Leave dishes undisturbed for 2 hrs. to allow the ether
   to disperse.

5. Seed each dish with approx. 100, 1 mm squares cut from
   the periphery of a young colony of *P. cinnamomi*, growin
   on cornmeal agar.

6. Incubate cultures at 20°C for 15 days under continuous
   fluorescent light (15 watts, daylight-type, at 20-50 cm
   above dishes).

7. Remove mycelial mats from the culture, wash three times

in sterile distilled water to remove any nutrients, and float in 125 ml of sterile distilled water in a stoppered separating funnel.

8. Shake funnel vigorously in a wrist-action flask shaker until the mycelium rises to the surface of the water (approx. 15 min.).

9. Strain contents of funnel through four layers of muslin into petri dishes.

10. Allow filtrate to settle for about 5 min.

11. Gently swirl the contents of the dish and collect the chlamydospores with a 1 ml pipette, and store in a test-tube.

12. Pipette off the supernatant from the test-tubes to obtain a highly concentrated chlamydospore suspension.

>te:

ıis technique significantly reduces chance contamination and ılamydospore damage when compared to other methods.

RLING, D.D. 1978. Phytophthora Newsletter 6: 8-9.

Production of chlamydospores in plant tissue: Lupin technique

1. Soak blue lupin seeds (*Lupinus angustifolius* L.) in 6% hydrogen peroxide for 5 hrs.

2. Then place the seeds on the surface of moist, sterile vermiculite in screw-top jars.

3. Lupin roots (3-5 cm long) are produced in 4 days.

4. Excise roots at the hypocotyl, cut into 1 cm lengths and place on the surface of a 3 day-old culture of *P. drechsleri* on V-8-oatmeal agar (10 ml V-8 juice, 20 g oatmeal, and 20 g agar per liter of distilled water).

5. Incubate for 6 days at $25^{\circ}C$, then remove the root pieces and bury them in moist field soil in 2.5 x 5.0 cm glass vials (4/vial).

6. Pack the vials in aluminum tins lined with moist paper towels, and seal the lids with adhesive tape to retain moisture.

7. Remove the lids every three days to facilitate aeration.

8. Remove the root pieces after three weeks and examine for chlamydospores by macerating in lactophenol/cotton blue.

9. This technique is good for chlamydospore production of *P. nicotianae* var. *parasitica*, *P. cinnamomi*, and *P. syringae*.

COTHER, E.J. and D.M GRIFFIN. 1973. Trans. Brit. Mycol. Soc. 61: 379-402.

H.   Production of chlamydospores of *P. cryptogea*

1. Place root pieces (about 15-20 mm long and 1 mm in diameter), of *Pinus radiata* on cultures of *P. cryptogea* and incubate at 25°C for 5-6 days.

2. Transfer colonized pieces to the surface of smoothed soil in small petri dishes and incubate at 20°C.

3. After 30 days, immerse root pieces in distilled water for 24 hrs.

4. Root pieces should now contain numerous chlamydospores which can easily be observed microscopically.

Note:

Germinated chlamydospores often have sporangia developing at the tips of the germ tubes.

BUMBIERIS, M. 1976. Phytophthora Newsletter 4: 16-17.

I. Chlamydospore formation: *P. nicotianae* var. *parasitica*   (*P. parasitica*)

Method I

1. Cut inoculum plugs of 7 mm diameter from the edge of a 5 to 10-day-old culture growing on cornmeal agar (Difco, 17 g/liter).

2. Place each plug in a 12 oz glass prescription bottle (vol. ca. 350 ml), containing 25 ml of a liquid medium.

3. The liquid medium consists of filtered V-8 juice, 100 ml; filtered 2% $CaCO_3$, 100 ml; and deionized water, 800 ml.

4. Incubate bottles at $25^\circ C$ in darkness for 22-24 hrs.

5. Pick up each bottle and shake vigorously (30 short strokes), to fragment the hyphae growing out from the inoculum plug.

6. Slowly rotate each bottle to resuspend hyphal fragments adhering to the walls of the bottle.

7. Incubate bottles horizontally at $25^\circ C$ for 6 days, so that the mycelium covers the surface of the medium.

8. Add 100 ml of sterile deionized water to each bottle.

9. Incubate, with the bottles standing vertically at $18^\circ C$. The mycelial mats should sink to the bottom of each bottle. If an occasional mycelial mat floats to the surface, gently shake the bottle to facilitate sinking.

10. After 2 to 3 weeks abundant chlamydospores (ca. 5×10^5 to 1×10^6 spores/bottle) are formed.

11. Harvest chlamydospores by filtering and washing mycelial mats on a Buchner funnel.

12. Blend mycelial mats for 3 minutes in 150 ml distilled water in a Sorvall Omni-mixer (setting of 88, surrounded by cold water to prevent overheating.

13. Partially separate mycelial fragments and chlamydospores by centrifuging suspension twice for 15 sec each in a clinical centrifuge.

14. Approximately 30-35% of the chlamydospores are lost du-

ring centrifugation.

15. Pool suspensions at the bottom of the centrifuge tubes
and wash by centrifugation three times for 3 min. at
1,500xg each time.

16. Determine the viability of harvested chlamydospores usin
60 μg/ml of rose bengal solution (see page 139).

Notes:

a)   It is important to incubate the cultures at 18°C; few
chlamydospores form at 27-30°C.

b)   Incubation temperature varies with the species.

TSAO, P.H. 1971. Phytopathology 61: 1412-1413.

Method II

1. Transfer 3 mm diameter discs to 250 ml Erlenmeyer flasks
containing clarified V-8 juice (10%) broth and incubate
in the dark at 25°C for 5 days.

2. Rinse the resulting mycelial mats aseptically twice with
distilled water.

3. Incubate the washed mats in 200 ml sterile distilled
water (approx. 6 cm deep) for 4 weeks at 16°C.

KUNIMOTO, R.K. *et al.* 1976. Phytopathology 66: 546-548.

Method III: Mycelium-free suspensions of chlamydospores

1. Take 21 day-old cultures of *P. nicotianae* var. *parasitica* growing in V-8 broth and comminute in a microblender
for 1 minute at maximum speed.

2. Homogenize the resulting suspension in a glass tissue
grinder.

3. Subject 50 ml aliquots of a suspension of the chlamydospores and mycelial fragments to maximum sonication usin
a Biosonik III (Bronwill Scientific Co.) ultrasonic system for up to 300 sec.

Notes:

a) Germination of chlamydospores is greater than 92% after 24
    hrs in water on dilution plates, following sonication
    treatment.

b) Sonication provides a rapid and reliable method of destroy-
    ing hyphal fragments, zoospores, and sporangia.

MITCHELL, D.J. and M.E. KANNWISCHER. 1976. Phytophthora Newsletter 4: 17-19.

J. Chlamydospore production: *P. palmivora*

1. Clarify a V-8 juice medium supplemented with 0.1% $CaCO_3$
    by centrifugation at 12,000xg for 10 min.

2. Dispense 25 ml of 20% clarified V-8 juice broth into
    each 250 ml flask and autoclave.

3. Inoculate each flask with 5 inoculum discs (6 mm diam),
    from a 4 day-old culture.

4. Incubate for 5 days at $22^{O}C$.

5. Replace the V-8 juice medium aseptically with 100 ml
    sterile distilled water.

6. Incubate the submerged cultures a further 4 weeks at
    $16^{O}C$ in darkness.

TSADOOKA, J.Y. and W.H. KO. 1973. Phytopathology 63: 559-562.

Section 6:

Sexual reproduction:

Oospore production and germination

I. Production of oospores of *Phytophthora* species

    1. Species of *Phytophthora* are either homothallic, hetero-
       thallic, or neuter.

    2. Homothallic species produce oospores in single culture
       on a suitable medium.

    3. Heterothallic species require two mating or compatibility
       types (A^1 and A^2) paired in the same culture medium   to
       form oospores, e.g. with *P. infestans*, isolates from
       Mexico may be either A^1 or A^2, whereas isolates from the
       U.S., Canada, Western Europe, South Africa, and the West
       Indies, comprise compatibility group A^1 only (Smoot *et
       al.* 1958. Phytopathology 48: 165-171, Galindo and Galleg-
       ly. 1960. Phytopathology 58: 1004-1021).

II. Distribution of mating or compatibility types

The geographical distribution of the A^1 and A^2 compatibility
types vary with different heterothallic species. For *P. cinnamomi* see Zentmyer, G.A. 1976. Phytopathology 66: 701-703. For
P. infestans see Gallegly, M.E. and A.J. Galindo, 1958. Phytopathology 48: 274-277.

For *P. palmivora* see Zentmyer, G.A. 1973. Phytopathology 63:
663-667.

Notes:

a) Sterols are essential for the formation of oospores. This
    is an absolute requirement when chemically defined media
    are used.

    The relative activity of sterols for formation of oospores
    are as follows: β-sitosterol = fucosterol = Δ^5- avenasterol
    = stigmasterol ⟩ cholesterol = 7-dehydrocholesterol ⟩ ergoste-
    rol ⟩ cholesterol (Elliot, C.G. 1972. J. Gen. Microbiol. 72:
    321-327). It is believed that sterols improve the integrity
    and structure of the plasma membrane.

See page 83 for concentrations and methods of adding
sterols to the medium.

b) Oospore production is greater at O_2 concentrations of 1%
   and 5% than in air, but decreases with increasing CO_2 concentrations when the O_2 level is 1, 5, or 20% (Mitchell and
   Zentmyer. Phytopathology 61: 807-812, 1971).

c) Oospores form easily when isolates of two different species
   are paired on the same culture medium (interspecific hybrid
   ization) e.g., *P. cinnamomi* (A^1) x *P. palmivora* (A^2), *P.
   capsici* (A^1) x *P. cambivora* (A^2) etc. See Table 8, page 237

d) Oospores of *P. palmivora* will form in detached leaves of
   Piper nigrum when inoculated with opposite mating types and
   incubated at $15-27.5^\circ C$ in darkness. Oospores do not form in
   light or in dark at $30^\circ C$ (Brasier, C.M. Trans. Brit. Mycol.
   Soc. 52: 273-279, 1969).

The various stages of sexual morphogensis in *P. capsici* are illustrated in Fig. 3, and the sequence of cell wall differentiation believed to occur during this morphogenesis is shown in
Fig. 4.

III. Factors influencing oospore production

1. Sterols: An absolute requirement in chemically defined
   media (see page 82).

2. Light: Oospores are generally inhibited by light. Oospor
   production is best in darkness (see page 163 for details

3. Temperature: Species differ in their temperature require
   ments for oospore production. In general, temperatures
   below the optimum for growth are conducive to oospore
   formation.

4. Nutrition: A carbon/nitrogen ratio of 94:1, (e.g. glucos
   5. g/liter and asparagine, 0.1 g/liter), enhances oospor
   production. Amino acids such as aspartic, glutamic and
   asparagine are best; valine and leucine are toxic.

5. Calcium: When added to synthetic liquid media, oospore
   production is greatly enhanced.

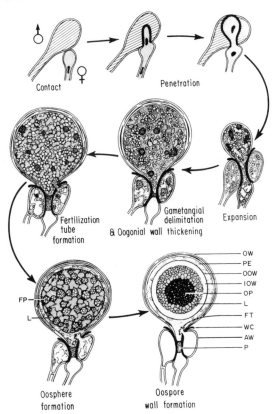

Fig. 3: Sexual morphogenesis in *Phytophthora capsici*: Gametangial interplay and oospore development. Antheridial wall (AW); "finger-print" vacuole (FP); fertilization tube (FT); inner oospore wall (IOW); lipid-like body (L); outer oospore wall (OOW); ooplast (Op); oogonial wall (OW); oogonial plug (P); periplasm (PE); wall collar (WC).

6. Depth of medium: Some species, e.g. *P. palmivora* form oospores in single culture when the depth of the lima bean agar medium in shallow (15-20 ml/90 mm petri plate), but not when the medium is deep (25-30 ml/90 mm petri plate).

Note:

A discussion on the control of sexuality in *P. cinnamomi* is given by Chang, S.T., C.J. Shepherd and B.H. Pratt. 1974. Austral. J. Botany 22: 669-679.

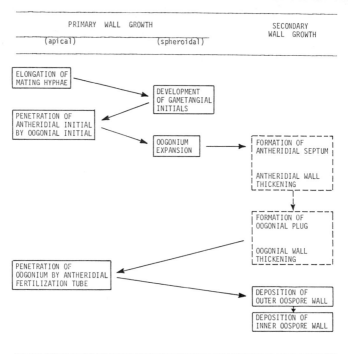

Fig. 4: Sequence of cell wall differentiation during sexual morphogenesis in *Phytophthora* (events listed inside dash-line rectangles are shown in their suspected sequence).

Figs. 3 and 4: Reproduced from HEMMES and BARTNICKI-GARCIA Archiv. Microbio. 103: 91-112. By permission of Springer-Verlag Copyright, 1975.

IV. Methods for the induction of oospores

1. In single culture (Homothallic species), or dual culture (Heterothallic species), oospores readily form in darkness on several natural and chemically defined media (see Media section). Savage *et al*. 1968. Phytopathology 58: 1004-1021.

2. Addition of β-sitosterol and tryptophan to natural media (see page 155). Chee *et al*. 1976. Plant Dis. Reptr. 60: 866-867.

3. By stimulation of certain species of *Trichoderma* (*T. Koningii, T. piluliferum, T. polysporum*, and *T. viride*)

heterothallic species of *Phytophthora* can be induced to form oospores (selfing). Brasier, C.M. 1975. New Phytol. 74: 183-194.

4. Oospores can be induced in single cultures of *P. cinnamomi* and *P. drechsleri* by the use of avocado root extracts. Zentmyer, G.A. 1952. Phytopathology 42: 24.

5. A^2 isolates of *P. capsici* form oospores in the presence of chloroneb (2,5-dichloro-1,4-dimethoxybenzene). Noon, J.P. and C.J. Hickman. 1974. Can. J. Botany 52: 1591-1595.

6. Stimulation of oospore production by damage or wounding of hyphae. Reeves, R.J. and R.M. Jackson. 1974. J. Gen. Microbiol. 84: 303-310.

7. By inoculation of natural hosts, e.g. *P. megasperma* forms abundant oospores in alfalfa root or stem pieces; *P. fragariae* forms oospores in strawberry roots, etc.

V. Oospore production: Tryptophan medium

1. Supplement clarified V-8 juice medium (page 75) with the following per liter of medium:

   β-sitosterol . . . . 30 mg

   Tryptophan . . . . 20 mg

   $CaCl_2.2H_2O$ . . . . 100 mg

   Thiamine . . . . 1 mg

2. Incubate in the dark at $20^\circ C$.

3. *Phytophthora* isolates that do not normally produce oospores in standard media (e.g. V-8 juice, oatmeal, cornmeal, etc.), have been successfully induced to form oospores on the above medium.

CHEE, K.H. *et al.* 1976. Plant Dis. Reptr. 60: 866-867.

VI. Production of Oospores in Liquid Medium

A. Make up V-8 CaCO$_3$ broth as follows:

1. Clarify V-8 juice by centrifuging at 6000 rpm (5860 x g)
   for 10 minutes, and filter the supernatant under suction
   three times through one layer, and two times through two
   layers of Whatman #42 filter paper.

2. Filter a 2% solution of the V-8 juice through one layer
   of filter paper to remove undissolved particles.

3. Combine 100 ml of filtered V-8 juice, 100 ml of filtered
   2% CaCO$_3$, and 800 ml of distilled water.

4. The final pH after autoclaving should be 4.3-4.5.

B. Prepare the inoculum as follows:

1. Take 7 mm plugs from the periphery of 7 day-old cultures
   growing on cornmeal agar plates and place in a 12 oz
   glass prescription bottle (vol. ca. 350 ml), containing
   Bartnicki-Garcia's medium (page 89), modified with 30 m
   ergosterol in 30 ml of dichloromethane.

2. Incubate inoculated bottles for 22-24 hrs are 25°C in
   darkness in a vertical position, after which shake each
   bottle 30 times to fragment hyphae.

3. Make sure that no hyphae adhere to the walls of the
   bottle to avoid formation of sporangia.

4. Then incubate the bottles in a horizontal position for
   6 days at 25°C in darkness.

5. Combine 3 cultures of an isolate in 100 ml of water and
   comminute for 10 secs in a Sorvall Omni-mixer.

6. Seed each 25 ml of V-8 CaCO$_3$ broth contained in a prescription bottle with 0.5 ml of the hyphal fragment suspension of each type (A^1 and A^2), in the case of heterothallic species or 1 ml in the case of homothallic species.

7. Incubate at 25°C in darkness.

8. *P. parasitica* oospores form in 3 days and continue for 10 days. (Approx. $1.5-3 \times 10^6$ oospores are formed/culture).

9. This method can also be used for oospore production of *P. palmivora*, *P. drechsleri*, *P. cinnamomi*, *P. megasperma* var. *sojae*, and *P. cactorum*. (Approx. $4 \times 10^4 - 9 \times 10^5$ oospores/culture are formed).

Note:

Temperature of incubation is important, e.g. for *P. parasitica* oospores form between $18-30^\circ C$, with an optimum between $21-24^\circ C$.

HONOUR, R. and P. TSAO. 1974. Mycologia 66: 1030-1038.

VII. Separation of oospores from mycelial mats

A. Oospore production

Method I

1. Place 1/2 of a 5 day-old *Phytophthora* culture (growing on V-8 juice agar in petri plate), into a Waring blender cup containing 100 ml of sterile distilled water. For heterothallic species use 1/2 plate each of the A^1 and A^2 mating types of the fungus. Comminute for 60 secs at 10 sec intervals to avoid heating.

2. Inoculate 15 ml of clarified V-8 juice broth (see page 75 for formula), in petri plates with the minced mycelium. Use 1 ml of the inoculum/plate.

3. Incubate in darkness for 30 days at $25^\circ C$.

Notes:

a) Wrap 1-2 plates in aluminum foil and monitor frequently to ascertain the quality and quantity of oospores being formed.

b) For some heterothallic species, better oospore production
   results if each mating type is comminuted separately and
   then 1 ml of each type added to the V-8 juice plates.

B. Separation of oospores from mycelial mats

   1. After 30 days, add 1 ml of a 0.1-0.5% solution of the
      snail gut enzyme (Helicase, see below), to each petri
      plate and incubate plates 24 hrs at 33°C.

   2. Comminute mycelial mats in a Waring blender for 2 min.
      at 10 sec intervals, to avoid heating. Add sterile dis-
      tilled water when necessary.

   3. Filter the minced mycelium through a sterile 53 μ mesh
      heavy duty nylon (Tobler, Ernst & Traber, Inc., 420 Saw
      Mill River Rd., Elmsford, N.Y. 10523, U.S.A.). This ny-
      lon can be repeatedly autoclaved. Support the nylon on
      a Buchner funnel mounted on a sterile 250 ml flask.

   4. Centrifuge filtrate in a clinical centrifuge, at full
      speed, for 1 min. Oospores collect at the bottom of the
      tube.

   5. Remove liquid and floating mycelial debris using a ster-
      ile Pasteur pipette attached to an aspirator.

   6. Add sterile deionized water and centrifuge again to re-
      move any remaining mycelial fragments.

Notes:

a) This method is useful for obtaining a large quantity of
   oospores rapidly under aseptic conditions. 100 petri plate
   cultures should yield 800-1000 mg of clean oospores.

b) Approximately 60% recovery of oospores is obtained with
   heterothallic species tested. Homothallic species, e.g. *P.
   megasperma* var. *sojae* produce oospores deep in the mycelia
   mats and separation of oospores from mycelial mats is difficult without repeated comminutions and centrifugations.

c) The enzyme Helicase fragments and destroys the mycelium,

facilitating oospore harvest. The enzyme can be obtained
from L'Industrie Biologique Francaise, 101 Ave de Verdun,
92 Villeneuve-La-Grarenne, Grenevilliers, France, or, Endo
Labs, Inc., Garden City, N.Y. 11350.

d) A mixture of Beta-glucuronidase and sulfatase, although less
effective, can also be used.

RIBEIRO, O.K. Dept. of Plant Pathology, University of California, Riverside, Ca. 92502, U.S.A.

Method II

1. Place portions of the agar culture in a beaker of water
with the water snail, *Planorbarius corneus* (Linn.).

2. Snails produce feces consisting almost entirely of
oospores.

3. After 48 hrs , collect the feces and suspend the oospores
in water with a glass tissue grinder.

4. Surface sterilize the oospores by adding to the suspension an equal volume of aqueous mercuric chloride
(60 mg/l) for 5 min.

5. Wash spores in four changes of sterile distilled water.

SHAW, D.S. 1967. Phytopathology 57: 454.

Note:

The garden snail, *Helix aspersa* has also been successfully used
for germination of oospores of *P. erythroseptica* (Gregg, M.
1957. Nature (Lond.) 180: 150).

Method III

1. Wash mycelial mats in distilled water.

2. Place in solution of 0.5%-2% mixture of cellulase and
hemicellulase in 0.1 M phosphate buffer at pH 6.0.

3. Incubate at 30°C, 2-7 days.

4. Remove bacteria and protoplasm from oospore suspension by washing through nylon net disc 25 mm diameter (20 μ pore size), using a Millipore apparatus.

5. Wash oospores from the nylon net by reversing the disc in the apparatus.

6. Centrifuge at 1000 g for 1 min. and resuspend in distilled water.

7. Aseptic oospores can be obtained by exposing them to 30 ppm $HgCl_2$ soln. for 5 min., followed by washings in sterile water.

SNEH, B. 1972. Can. J. Bot. 50: 2685-2686.

Method IV

1. Sonicate mycelial mats to fragment mycelium and free oospores.

2. Kill mycelial fragments and sporangia by freezing at $-20^{O}C$ for 24 hrs.

BANIHASHEMI, Z. and J.E. MITCHELL. 1976. Phytopathology 66: 443-448.

Method V

Separation of oospores of P. megasporum var. sojae from mycelia mats.

1. Oospores are produced in cultures of P. megasperma var. sojae growing in 1-liter Roux flasks containing 100 ml of a centrifuged cleared V-8 juice-$CaCO_3$ broth at $22-24^{O}C$.

2. Wash mycelial mats and blend for 8 min. in 0.1 M sucrose in a Sorvall Omni-mixer at 6000 rpm.

3. Place blended cultures on a reciprocal shaker at 130 excursions per min. for 2 hrs.

4. Remove from shaker, centrifuge, suspend in 1 M sucrose, and freeze for 12 hrs to kill mycelial fragments.

5. Rewash and resuspend in deionized distilled water.

6. Layer over 1 M sucrose and centrifuge for 1 min. at 1,500xg.

7. Oospores collect at the bottom of the tube.

Note:

Approx. 500 mg of oospores are collected from 20 flasks.

PARTRIDGE, J.E. and D.C. ERWIN. 1969. Phytopathology 59: 14 (Abstr.).

VIII. Oospore dormancy

Phytophthora oospores are believed to be constitutively dormant, i.e. development is delayed due to an intrinsic property of the spore such as resistance to the penetration of nutrients, a metabolic block, or a self-inhibitor. There is thus far no evidence to support any one of these alternatives as the cause for dormancy in *Phytophthora* oospores.

Useful References:

COCHRANE, V.W. 1974. Dormancy in spores of fungi. Trans. Amer. Microscopic Soc. 93: 599-609.

SUSSMAN, A.S. and H.A. DOUTHIT. 1973. Dormancy in microbial spores. Annu. Rev. Plant Physiol. 24: 311-352.

Note:

It is important to distinguish between immaturity and dormancy, since oogonial (immature) germination is known to occur in *Phytophthora*. *Phytophthora* oospores require a period of at least 10 days to mature (Blackwell, Trans. Brit. Mycol. Soc. 26: 93-103, 1943).

IX. Test for viability of oospores

Method

1. Make up a 0.1% solution of 3(4,5-dimethyl thiazolyl 1-2)
   2,5 diphenyl tetrazolium bromide (MTT) in 0.1 M phosphate
   buffer, pH 5.8.

2. Add 0.5 ml 0.1% MTT to 0.5 ml of oospores in a vial.

3. Incubate in a water bath at 35°C for 2-2 1/2 hrs.

4. Examine microscopically.

Notes:

a) Color of viable oospores range from a dark blue to a deep
    red, depending on species used. Non-viable oospores do not
    stain.

b) Other tetrazolium compounds are not as effective.

RIBEIRO, O.K. 1973. Dept. of Plant Pathology, Univ. of California, Riverside

B. Germination of oospores

The germination of oospores of *Phytophthora* species in artificial culture has long been a barrier to meaningful genetic stud
ies on inheritance mechanisms involving pathogenicity, compatibility types, morphology, etc. Consistent germination still remains elusive for most species.

The germination process is eloquently described by Erwin and
McCormick (Mycologia 63: 972-977, 1971).

I. The following factors are believed to be involved in the
    breaking of dormancy of *Phytophthora* oospores.

    1. Maturity

       Blackwell (1943) believed that oospores required a 'rest

period' of 1-2 months prior to germination. Satour and
Butler (1968), also indicated that maturity was a fac-
tor in germination. They found that young oospores of *P.
capsici* gave low germination (1.9-2.6%), compared to 60
day-old oospores that gave 19-20% germination. However,
MacIntyre and Elliott (1973) found that young (10 day-
old) and older (21 day-old) oospores of *P. cactorum* germinated equally well. Ribeiro (1976) also found no difference between the germination of young and older
oospores (up to 305 days), of *P. capsici* and *P. cinnamomi*.

Boccas (1970), and others believe that oospores have the
ability to germinate immediately after formation. Under
unfavorable conditions, the oospores do not germinate,
but go into a period of dormancy.

2. Light

Berg and Gallegly (1966), reported that 70% of the
oospores of *P. cactorum* germinated when placed under
blue light, compared to 36% under far-red light (750 nm),
10% under normal white light, and only 5% in darkness.
These results were confirmed by Huguenin and Boccas
(1971). They found that radiation below 500 nm enhanced
germination of *P. palmivora*. Ribeiro *et al.* (1975 & 1976)
confirmed the enhancement of oospore germination by blue
and far-red radiation. They also found that the duration
of exposure to light was not important for germination.
For maximum effectiveness, irradiation must be carried
out on mature oospores. Blue and far-red radiation have
little effect on oospore germination if the exposure
occurs during gametogenesis.

3. Temperature

Conflicting reports exist on the importance of temperature on oospore germination. Some investigators have
found that low temperature treatments enhance the break-
ing of dormancy and subsequent germination, while others
have not obtained any stimulation with low temperatures.
Long *et al.* (1975) obtained significant stimulation of

oospore germination of *P. megasperma* var. *sojae* at 36°C
for 48 hrs , while Klisiewicz (1970) found that oospores
of *P. drechsleri* germinated in a wide range of temperatures (15-30°C).

4. pH

Oospores germinate in a large pH range. Reports indicate
that *P. cactorum* can germinate in a pH range from 7.6-
10; *P. capsici*, pH 4.5-7.3; *P. drechsleri*, pH 4.5-9.5;
and *P. infestans*, pH 2-10. Huguenin and Boccas (1971) re
ported a narrow range of pH 5-6 for optimum germination
of *P. palmivora*.

5. Snail gut enzyme

Passage of oospores through snails have resulted in high
oospore germination of *P. cactorum, P. erythroseptica*,
and *P. megasperma* (see pages 167-169). The snail species
used were *Planorbarius corneus* (Linn.), *Helix aspersa*
(Muller), and *Eopania vermiculata* (Muller). Similar results have been reported using the snail gut enzyme,
Helicase (see page 158). A mixture of B-glucuronidase
and aryl sulfatase can also be used, but is less effective than Helicase. The advantage of this mixture is
that it is readily available and less expensive than
Helicase. Concentrations of the enzyme used for oospore
germination range from 0.1% to 5.0% for 24-48 hours.
However, it should be noted that Long *et al.* (1975) foun
that Helicase (0.025%-5%), for 1-72 hrs at 25°C, 30°C
and 36°C did not stimulate oospore germination of *P. megasperma* var. *sojae*, but rather in some instances reduce
germination when compared to water-treated controls.

6. Permeability

The thick oospore wall appears to be permeable to most
chemicals, since cytological and vital dyes are able to
penetrate and stain the oospore. The oospore wall is
thus not believed to be the major barrier to germinatior
The structure of the oogonial-oospore wall complex is
described on page 271.

7. Oospore contents

Leary *et al.* (1974), did not detect the presence of intact ribosomes in oospores. Ribosomal RNA and protein
however, were found to be present in oospores. Ribeiro
et al. (1976) could detect cytochrome C in mature
oospores, but cytochromes a and b were found to be lacking, or in quantities too minute to be easily detected.

Size and visual appearance of oospores also have no effect on ability to germinate.

8. Genetic proneness

Several investigators have found that some isolates and
parental crosses (A^1 x A^2) of *Phytophthora* are better
able to germinate than others. Romero and Erwin (1969)
obtained oospore germination ranging from 64% for one
cross of *P. infestans*, to 6.5% for another cross involving isolates obtained from the same host and location.

9. Miscellaneous techniques

Most of the classical methods listed by Sussman and Halvorson (1966) have been tried to induce oospore germination. Other methods include: influence of various growth
media, e.g. PDA, V-8 juice, oatmeal, etc.; potato soil
leachate; horse dung infusion; heat shock; alternate
wetting and drying; freezing and thawing; UV radiation;
growth promoting hormones; volatile substances; weak
acids; amino acids, vitamins; Ca++ ions, Mg+ ions; etc.

Notes:

Our experience seems to indicate that oospore germination   is
largely a matter of trial and error in trying to find the precise conditions for the particular isolate being studied. Even
isolates of the same species differ in their requirements   for
oospore germination and as a result, no one method of oospore
germination has found to be universally applicable.

References:

BERG, L.A. and M.E. GALLEGLY. 1966. Phytopathology 56: 583.

BLACKWELL, E. 1943. Trans. Brit. Mycol. Soc. 26: 93-103.

BOCCAS, B. 1970. C.R. Acad. Sci., Paris. 270: 55-58.

HUGUENIN, B. and B. BOCCAS. 1971. Annales Phytopathol., Paris. 3: 353-371.

KLISIEWICZ, J.M. 1970. Phytopathology 70: 1783-1742.

KOUYEAS, H. 1975. Phytophthora Newsletter 3: 7-9.

LEARY, J.V. et al. 1974. Phytopathology 64: 404-408.

LONG et al. 1975. Phytopathology 65: 592-597.

MacINTYRE, D. and C.G. ELLIOTT. 1973. Phytophthora Newsletter 1: 20-21.

RIBEIRO, O.K. et al. 1975. Phytopathology 65: 904-907.

RIBEIRO, O.K. et al. 1976. Phytopathology 66: 172-174.

RIBEIRO, O.K. et al. 1976. Phytopathology 68: 1162-1173.

RIBEIRO, O.K. et al. 1976. Mycologia 68: 1162-1173.

ROMERO, S. and D.C. ERWIN. 1969. Phytopathology 59: 1310-1317.

SUSSMAN, A.S. and H.O. HALVORSON. 1966. Spores, their Dormancy and Germination. Harper & Row, New York, 354 p.

Species	Medium	after oospore harvest	(%)	Age of Culture (days)	Authority
P. cactorum	not given	Chill oospores 1-2 wk then H$_2$O soak (pH 7.6-8.0) for 1-3 wk.	not given	30+	Blackwell (1943)
P. erythroseptica	oat agar	Ingestion by garden snail, Helix aspersa.	not given	30-60	Gregg (1957)
P. infestans	soybean meal, V-8J lima bean	Horse dung infusion. Potato soil leachate.	3-4	not given	Smoot et al. (1958)
P. heveae	synthetic	25 days dark, 12-15 days light	65-70	39+	Leal & Gomez-Miranda (1965)
P. cactorum P. heveae P. erythroseptica P. infestans	lima bean	19 days dark. Fluorescent light (500-700 ft c.) for 7 days	90+ 75+ 50+ 10+	26+	Berg (1966)
P. cactorum	sucrose-asparagine & ether extr. oats + pea infusion	Ingestion by snail, Planorbarius corneus, followed by 4-8 days at 18 C-22 C in diffuse light.	67	21+	Shaw (1967)
P. citricola	lima bean	21-25 C fluorescent light + moderate sunlight-unregulated	50	not given	Henry & Stelfox (1968)
P. capsici	V-8J Rape Seed	Mixture of daylight & fluorescent light (160 ft c.).	15-20	45-60	Satour & Butler (1968)
P. infestans	V-8J	Clarified V-8J + beta-sitosterol	20-48	20+	Romero (1969

Table 6b: Methods to induce oospore germination

Species	Culture Medium	Treatment to induce germination	Germination (%)	Age of Culture (days)	Authority
P. drechsleri	cornmeal safflower	Sterile tap H_2O (pH 7.5) at 24 C exposed to white fluorescent light (45 ft c.) 10 hr, dark 14 hr.	40-53	35-56	Klisiewicz (1970)
P. megasperma var. sojae	V-8J	2 hr/day fluorescent light (200 ft c.) for 7 days. Frozen 24 hr. Incubated in H_2O 4 days.	40-70	10-20 +30 at 9C	Erwin & McCormick (1971)
P. capsici	V-8J	Lab light & temp.	6-8	60+	Polach & Webster (1972)
P. cactorum	pea extract	Continuous light for 14 days at 25C.	74.3	21+	Elliott & MacIntyre (1973)
P. megasperma	V-8J	Injestion by snails.	93	35+	Salvatore et al. (1973)
P. megasperma	V-8J	2000 units/ml of β-glucuronidase for 144 hr.	52.3	35+	Salvatore et al. (1973)
P. cinnamomi	V-8J broth	Dark 10 days/blue light 10 days/ dark 10 days.	23.0	30	Ribeiro (1974)
P. megasperma var. sojae	V-8J agar	Water soak 48 hr at 36C. Overlay on water agar. Lab. light and temp. (24C)	45-67	21+	Long (1974)
P. drechsleri	synthetic V-8J agar	Overlay on water agar. Lab light and temp.	39.0 42.0	30	Ribeiro (1974)

Table 6c: Methods to induce oospore germination

Species	Culture Medium	Treatment to induce germination	Germination (%)	Age of Culture (days)	Authority
P. cactorum	V-8J broth	Overlay frozen oospores on 2% agar after treatment with 2% glusulase. Expose to 2,165 lux continuous cool white fluorescent light for 7 days at 22 C.	76%	120	Banihashemi & Mitchell (1976)
P. cactorum	V-8J broth	Overlay oospores on 2% agar and expose to 2594 lux continuous cool-white light for 104 hr.	91%	120	Banihashemi & Mitchell (1976)
P. fragariae	Roots of fragaria	Surface sterilize 2 min. in 0.001% mercuric chloride, wash 10-15 times in distilled water. Incubate on 1% water agar 2-3 wks.	5-60%	28-30	Duncan (1977)

Section 7:

Penetration, dissemination, and survival

I. Mode of Penetration

Phytophthora zoospores usually penetrate directly through roots or through stomata, depending on whether the species is soil-borne or aerially dispersed.

Zoospores are attracted to, and accumulate on roots by chemotaxis and/or electrotaxis (see pages 296-298). In the case of aerial borne inoculum, e.g. *P. infestans*, zoospores after encystment on the leaf surface, germinate by producing germ tubes which develop appressoria (a flat hyphal organ) which press against the epidermal cells of the host. A minute infection peg then grows out of the appressorium and penetrates the epidermal cell of the host. Penetration is partially the result of enzyme action.

I. Haustoria

Haustoria (absorbing organs originating on the hyphae of the parasite that derive food from the host by penetrating the cells), have been described for some *Phytophthora* species. The shape of the haustorium varies greatly. In *P. infestans*, the haustorium is club-shaped, curled, or spirally twisted. Its diameter rarely exceeds 1 or 2 μm. Haustoria of *P. infestans* can develop on mycelium penetrating the tissues of leaf, aerial stem, or tuber of the potato. Haustoria have also been described in *P. arecae* (fingerlike, sometimes dicotomous); *P. boehmeriae* (cylindrical); *P. cactorum* (spherical); *P. cambivora* (globose or filamentous); *P. colocasiae* (?); *P. cyperi* (ovoid); *P. nicotianae* var. *parasitica* (cylindrical, finger-shaped or button-shaped, rarely branched); *P. palmivora* (branched, finger-shaped); and *P. phaseoli* (rare, irregular).

BLACKWELL, E. 1953. Haustoria of *Phytophthora infestans* and some other species. Trans. Brit. Mycol. Soc. 36: 138-158.

EHRLICH, M.A. and H.G. EHRLICH. 1966. Ultrastructure of the hyphae and haustoria of *Phytophthora infestans* and hyphae of *P. parasitica*. Can. J. Botany 44: 1495-1503.

174

III. Infective propagules

Sporangia of *P. palmivora* are reported to be the most infective propagules to papaya seedlings; chlamydospores were intermediate; zoospores the least infective (Ko and Chan, Phytopathology 64: 1307-1309, 1974). With *P. megasperma* var. *sojae* on soybeans and *P. megasperma* on alfalfa, mycelial fragments have been found to be effective for infection. Under appropriate conditions of humidity and temperature, zoospores of *P. infestans* and *P. palmivora* have been found to be effective in infecting their respective hosts, potato and cacao. However, it has been found that with cacao pods, mycelial fragments are often more effective than zoospores in initiating infection.

IV. Dissemination

Phytophthora inoculum is dispersed in one or more of the following methods.

1. Sporangia dispersed by wind and rain.
2. Plant debris, volunteer growth, cull piles, etc.
3. Infected seeds, pods, cuttings, etc.
4. Surface water and run-off.
5. Susceptible weed hosts.
6. Contaminated soil, implements, vehicles (e.g. logging trucks).
7. Insects.
8. Animals, e.g. through feces of snails, rodents, etc.
9. Workers transporting infested soil on shoes and clothing

V. Survival

Phytophthora species are not considered to be strong saprophytes and rarely survive in soil as mycelium if hosts or plant debris are removed from the field. Mycelium often survives by growing

from year to year in living host tissues, or in crop residues.
However, Newhook (1959) reported that *Phytophthora* has been
found to persist in soil for at least two years after the death
of trees in conifer forests, while Zentmyer and Erwin (1970)
found that *P. cinnamomi* persisted up to 6 years in the absence
of a host when naturally-infested soil was kept moist at 20°C.
The survival rate dropped to only a few weeks when the soil was
allowed to dry to 2-3% moisture content. Other investigators
have found that *Phytophthora* mycelium cannot survive over winter
as a saprophyte.

Honour (1973) found that soil *Streptomyces* species lysed mycelium of *P. parasitica* in 4 days and complete dissolution of the
mycelium occurred in 10 days. Oogonia were found to be more resistant to lysis by soil *Streptomyces* and other microorganisms
than hyphae or antheridia. Oogonia became melanized in soil and
it is believed that this further increased the resistance of
oospores to lysis by microorganisms in the soil. Intact oospores
at 60 days had 41.6% melanized type compared to only 14.3% of
the non-pigmented type. The presence of oogonia and intact
oospores were found to decline after 20 days.

Sneh *et al.* (1977), found that oospores of *P. megasperma* var.
sojae and *P. cactorum* were parasitized by *Pythium* sp.; *Leptoge-
ria* sp.. Hyphomycetes such as *Dactylella spermatophaga; Dihete-
rospora chlamydosporia; Humicola fuscoatra; Fusarium oxysporum;
Cephalosporium sp.*, and *Alternaria alternata*. Chytrids such as
Rhizodiomycopsis japnicus, Canteriomyces stigcoelonii, and *Hypho-
chytrium catenoides*. Actinomycetes such as *Actinoplanes mis-
souriensis*, and bacteria such as *Pseudomonas* spp.

nfection occurred within one week and by three weeks, 60-80%
of the oospores were infected.

On the other hand, Hickman (1958) reported that oospores of *P.
fragariae* survived 1 1/2 yr in non-sterile soil.

There are conflicting reports on the value of chlamydospores as
survival structures in the life cycle of fungi. Zentmyer and
Erwin (1970) claimed that the chlamydospore is a primary unit

of survival for *Phytophthora* while Chee (1973) found that
chlamydospores of *P. palmivora* did not survive long enough in
soil to be considered of importance in long term survival of
the fungus.

CHEE, K.H. 1973. Trans. Brit. Mycol. Soc. 61: 21-26.

HICKMAN, C.J. 1958. Trans. Brit. Mycol. Soc. 41: 1-13.

HONOUR, R.C. 1973. Ph.D. Dissertation, Univer. of Calif., Riverside, Ca.

MARKS, G.C. and J.E. MITCHELL. 1971. Can. J. Botany 49: 63-67.

SNEH, B. *et al*. 1977. Phytopathology 67: 622-628.

ZENTMYER, G.A. and D.C. ERWIN. 1970. Phytopathology 60: 1120-1127.

Section 8:

Pathogenicity tests

8A Pathogenicity to seedlings

and fruits

Although most *Phytophthora* species cannot be readily distinguished by their pathogenicity to various plants and fruits,
the following tests are often helpful in separating pathogenic
from non-pathogenic species and strains to a particular host.

Pathogenicity to various seedlings and fruits

I. Pathogenicity to apples

  1. Place mycelium or spores from cultures on agar or in
     tissue in small slits about 3 mm deep.

  2. Cover wound with Vaseline.

  3. Incubate at room temperature.

Symptoms:

  1. Infected apples become slightly brown on the exterior.

  2. No soft wet rot occurs.

Species tested:

Virulently pathogenic	Weakly pathogenic	Nonpathogenic
P. arecae	*P. cambivora*	*P. colocasiae*
P. boehmeriae	*P. cryptogea*	*P. infestans*
P. cactorum	*P. erythroseptica*	*P. phaseoli*
P. capsici	*P. drechsleri*	*P. richardiae*
P. cinnamomi		*P. megasperma*
P. citricola		var. *megasperma*
P. citrophthora		
P. hibernalis		
P. meadii		
P. mexicana		
P. nicotianae var. *parasitica*		
P. palmivora		
P. syringae		

Notes:

a) Most species grow rapidly in apple tissues and hence isola-
    tion of the fungus is quite successful with this technique.

b) This method is particularly useful for isolations from roots
    where saprophytes (e.g. *Fusarium* species and bacteria) are
    numerous.

II. Pathogenicity to castor bean (*Ricinus*)

    1. Inoculate castor bean seedlings by placing the inoculum
       in a small slit in the stem about 4 cm below the tip.

    2. Incubate in the greenhouse.

    Symptoms:

    1. Pathogenic species cause dark brown to nearly black
       stem lesions. Parts above the wound wilt and die.

    Species tested:

Virulently pathogenic	Nonpathogenic
P. capsici	*P. arecae*
P. meadii	*P. boehmeriae*
P. nicotianae **var.** *parasitica*	*P. cactorum*
P. palmivora	*P. cambivora*
	P. cinnamomi
	P. citricola
	P. citrophthora
	P. colocasiae
	P. cryptogea
	P. drechsleri
	P. erythroseptica
	P. mexicana
	P. richardiae

III. Pathogenicity to eggplant (*Solanum melongena* L.) fruit

   Method: Same as for tomato fruits (see page 184).

   Species tested:

Virulently pathogenic	Very weakly pathogenic	Nonpathogenic
P. cactorum	P. cactorum	P. arecae
P. capsici	P. palmivora	P. boehmeriae
P. citrophthora		P. cambivora
P. cryptogea		P. cinnamomi
P. drechsleri		P. citricola
P. infestans		P. colocasiae
P. mexicana		P. erythroseptica
P. nicotianae var. parasitca		P. hibernalis
P. palmivora		P. meadii
		P. richardiae
		P. syringae

Notes:

Same as for tomato fruits (see page 184). Some isolates of *P. palmivora* may be weakly pathogenic or nonpathogenic.

IV. Pathogenicity to eggplant (*Solanum melongena* L.) seedlings

   Method: Wound stems of eggplant seedlings and inoculate about midway between the base and tip.

   Symptoms: Infection causes dark, sunken areas and results in death when the invaded region girdles the stem.

   Species tested:

Virulently pathogenic	Weakly pathogenic	Nonpathogenic
P. capsici	P. cactorum	P. arecae
P. mexicana	P. citrophthora	P. boehmeriae
P. nicotianae var. parasitica	P. meadii	P. cambivora
P. palmivora	P. drechsleri	P. cinnamomi
		P. citricola

Virulently pathogenic	Weakly pathogenic	Nonpathogenic
		P. colocasiae
		P. cryptogea
		P. erythroseptica
		P. richardiae

Note:

Some isolates of *P. palmivora* may be weakly pathogenic or non-pathogenic.

V. Pathogenicity to pawpaw (*Carica*) seedlings

1. Inoculate young stems midway between the base and the tip.

Symptoms:

A soft, wet, green to black rot followed by collapse of the stems.

Species tested:

Virulently pathogenic	Nonpathogenic
P. cactorum	*P. arecae*
P. capsici	*P. boehmeriae*
P. mexicana	*P. cambivora*
P. nicotianae var. *parasitica*	*P. cinnamomi*
P. palmivora	*P. citricola*
P. parasitica	*P. citrophthora*
	P. colocasiae
	P. cryptogea
	P. erythroseptica
	P. meadii
	P. richardiae

VI. Pathogenicity to potato tubers

1. Make a narrow slit into the potato tuber.

2. Insert mycelium or spores of the fungus into the wounds, and cover with tape.

3. Incubate at room temperature.

Symptoms: Cut surfaces of infected tubers are usually white, turning pink on exposure to air and finally turn a dark brown after 2 hrs.

Species tested:

Pathogenic	Nonpathogenic
P. arecae	*P. boehmeriae*
P. cactorum	*P. cambivora*
P. capsici	*P. citricola*
P. cinnamomi	*P. colocasiae*
P. citrophthora	*P. hibernalis*
P. cryptogea	*P. phaseoli*
P. drechsleri	*P. richardiae*
P. erythroseptica	*P. syringae*
P. infestans	
P. meadii	
P. mexicana	
P. nicotianae var. *parasitica*	
P. palmivora	

Notes:

pink coloration is characteristic of most pathogenic species xcept *P. infestans*, which causes a brown rot. Some isolates of . *palmivora* are nonpathogenic.

II. Pathogenicity to tomato seedlings

1. Wound tomato seedlings and inoculate 1/2 in above the surface of soil.

Symptoms:

1. In young seedlings, symptoms are a soft, wet rot similar to damping off.

2. In older plants, symptoms consist of a brown, sunken lesion.

Species tested:

Virulently pathogenic	Weakly pathogenic	Nonpathogenic
P. *capsici*	P. *boehmeriae*	P. *arecae*
P. *cryptogea*	P. *cactorum*	P. *cambivora*
P. *drechsleri*	P. *mexicana*	P. *cinnamomi*
P. *nicotianae* var.		P. *citricola*
parasitica		P. *citrophthora*
P. *palmivora*		P. *colocasiae*
		P. *erythroseptica*
		P. *meadii*
		P. *richardiae*

Notes:

Rotting is greatest when the plants are young. Some isolates of P. *nicotianae* var. *parasitica* and P. *palmivora,* are nonpathogenic.

VIII. Pathogenicity to tomato fruits

1. Wound a green tomato fruit with a sterile scalpel.

2. Insert mycelium or spores into the wound and cover with tape.

3. Incubate at room temperature.

Species tested:

Virulently pathogenic	Very weakly pathogenic	Nonpathogenic
P. *cactorum*	P. *boehmeriae*	P. *arecae*
P. *capsici*	P. *cinnamomi*	P. *cambivora*

	Very weakly	
Virulently pathogenic	pathogenic	Nonpathogenic
P. *citrophthora*	P. *citricola*	P. *colocasiae*
P. *cryptogea*	P. *drechsleri*	P. *richardiae*
P. *erythroseptica*		
P. *meadii*		
P. *mexicana*		
P. *nicotianae* var. *parasitica*		
P. *palmivora*		
P. *syringae*		

Notes:

Different isolations of the same species can show marked variations in pathogenicity. The failure of some isolates to infect is not attributed to a loss of virulence in culture, but rather to an inherent characteristic of each particular strain.

Some isolates of P. *nicotianae* var. *parasitica*, P. *cinnamomi*, and P. *palmivora* are nonpathogenic.

Methods I-VIII from: TUCKER, C.M. 1931. Mo. Agr. Exp. Sta. Bull. 153: 208.

Pathogenicity Tests

8B: Techniques for soil infestation and

other pathogenicity studies

Mycelial fragments, sporangia, zoospores, chlamydospores, and oospores have all been used successfully in pathogenicity tests. Although reports indicate that some propagules are better than others for initiating infection, the choice of inoculum often depends on the species. Some species do not form sporangia, chlamydospores or oospores easily, or in sufficient quantities for use in pathogenicity tests, thus limiting the choice of inoculum to mycelium or mycelial fragments. The major disadvantage of this type of inoculum is that it is difficult to accurately estimate the number of infective propagules. This is not the case if sporangia, zoospores, chlamydospores or oospores are used as inoculum.

The number of propagules can be measured by one of several methods, e.g. Haemocytometer, Levy-Hauser counting chamber, eel worm counter, microsyringe, or coulter counter.

The viability of chlamydospores can be ascertained by using rose bengal (see page 139). The criteria of viability is based on the theory that injured or non-viable propagules are selectively stained. MTT (see page 290), selectively stains viable oospores.

. Preparation of inoculum for soil infestation studies: *P. nicotianae* var. *parasitica*, (*P. parasitica*)

1. Seed each 100 ml of potato-dextrose broth (per liter extract from 250 g potatoes; dextrose 20 g), in 32-ounce prescription bottles, with a 7 mm agar mycelium plug. Cut the plugs from the edge of 7 to 15 day-old fungus cultures growing on potato-dextrose agar plates, with the aid of a cork-borer.

2. Incubate the bottles standing them vertically, at $25 \pm 2^{\circ}$C for 22-24 hrs.

3. Then place the bottles on a reciprocal shaker for 1.5-2 hrs to break up the hyphae growing out from the mycelial plugs.

4. Remove the bottles from the shaker and incubate in a horizontal position. In 7 days, the mycelium should

nearly cover the entire surface of the medium.

5. Harvest the mycelial mats from the bottle, pool together, and filter through 4 layers of cheesecloth mounted over a Buchner funnel. Rinse repeatedly with deionized water to remove any remaining nutrients.

6. Comminute mycelial mats in a Waring blender for 10-20 secs, using small aliquots of water.

7. Mix the mycelial fragment suspension thoroughly with a calculated amount of a sterile coarse sand and use a measured amount of the sand-inoculum mix to infest the soil.

Note:

This method of inoculation is suitable for *in situ* studies on the influence of soil temperature, soil moisture, soil type, soil microbial antagonists etc.

TSAO, P.H. and M.J. GARBER. 1960. Plant Dis. Reptr. 44: 710-715.

II. Periodic waterlogging for greenhouse *in situ* soil infestation studies

1. Place a clay saucer at the bottom of each pot and flood for 3 days with water in addition to watering the soil surface.

2. For the next 4 days, invert the saucer and water from the top, only when the surface appears dry.

3. Repeat this weekly cycle of 3 days flooding and 4 days normal watering throughout the duration of the experiment.

4. This method provides optimal conditions for sporangia production and zoospore liberation in repeated weekly cycles.

Notes:

a) The use of saucers prevents the dissemination of *Phytophthora* to adjacent pots.

b) Prolonged periodic waterlogging of the soil can cause substantial leaching and loss of certain soluble chemicals in the soil.

c) This method has been successfully used for pathogenicity tests with *P. nicotianae* var. *parasitica* to citrus and *P. megasperma* to alfalfa.

d) Regular watering methods result in poor root infection in infested soils. Continuous waterlogging results in severe root infection, but roots of control plants show signs of asphyxiation and are greatly reduced. The periodic waterlogging method described above appears to be the most satisfactory since roots of non-inoculated plants subjected to this method appear normal.

SAO, P.H. and M.J. GARBER. 1960. Plant Dis. Reptr. 44: 710-715.

II. Assessing degree of root infection: Healty root tip count method

1. Harvest seedling roots individually by gently removing the plant and soil from the pots containing infested soil.

2. Remove adhering soil particles from the roots by a jet of water spray.

3. Spread root systems on a piece of dark-colored paper and count the numbers of healthy-looking white root tips on each seedling.

4. Compare with numbers obtained from root systems in non-infested soil.

Note:

This method is useful for experiments of short durations, ranging from 4 to 6 weeks.

TSAO, P.H. and M.J. GARBER, 1960. Plant Dis. Reptr. 44: 710-715.

IV. Estimating resistance or susceptibility of citrus fibrous
    roots

    Method I

    1. Fasten seedlings to paraffined block of wood or corks
       with a rubber band and float on aerated water so that
       only the roots are immersed.

    2. Add zoospores to water and estimate percentage decayed
       roots after one week.

    Method II

    1. Dip or expose for several days the root systems of vari
       ties under test in a zoospore suspension.

    2. Plant in soil along with a similarly inoculated suscep-
       tible seedling for comparison.

    3. After 4-6 weeks in pots in soil, wash out the root sys-
       tems and estimate the percentage of rotted roots.

    Method III

    1. After inoculating the seedlings, plant them in a peat
       moss-vermiculite mixture.

    2. Incubate for two months.

    3. Estimate resistance on the basis of percentage survivor

KLOTZ, L.J. and T.A. DeWOLFE. 1960. Plant Dis. Reptr. 44: 572-573.

V. Pathogenicity tests for screening soybean varieties: *P. megasperma* **var.** *sojae*

Method I

1. Germinate soybean seeds in greenhouse pans containing a mixture of 50% peat moss overlaid with vermiculite.

2. When seedlings are 5 to 14 days old, make a 1-2 cm slit on the hypocotyls with a No. 20 gauge hypodermic needle, and simultaneously introduce into the wound about 0.05 ml of minced mycelium from a syringe.

3. Maintain plants at 100% relative humidity for 24-96 hrs.

4. Susceptible varieties will show rotting of the stems, resistant varieties exhibit a pink discoloration in the vicinity of the wound.

KEEN, N.T. *et al*. 1971. Phytopathology 61: 1084-1089.

Method II

1. Transfer pieces of mycelium of *P. megasperma* **var.** *sojae* from slants of lima bean agar to petri dishes containing 10 ml of the same medium at 1/3 of its original concentration.

2. Transfer a piece of this inoculum to the center of a 90 mm petri dish.

3. The inoculum should reach the edge of the plate in 7-10 days at room temperature. This inoculum usually suffices for over 100 seedlings.

4. Scrape the entire contents of the dish into a small beaker and add 5 ml distilled water. Stir vigorously until the slurry obtained can be easily drawn up into a 2-5 cc syringe fitted with a No. 16 gauge needle.

5. Place soybean seeds to be tested on moist towels.

6. Roll the towels and place in a germinator at 25°C.

7. Four days later, remove ungerminated seeds and abnormal seedlings.

8. Align the rest of the seedlings with the cotyledons

near the top edge of the towel. Keep seedlings moist at all times.

9. Make a 1-2 cm slit near the base of the hypocotyl, and fill with inoculum using a syringe (approx. 1 ml/seedling).

10. Place seedlings at 25°C.

11. Four days after inoculation, a slight pink discoloration of the wound takes place on a resistant variety, while suspectible varieties develop a brown, soft decay with subsequent death of the seedling.

Note:

An incubation temperature of 30°C will produce symptoms in most resistant varieties.

SCHOEN, J.F. 1967. Assoc. Official Seed Analysts, Proc. 67: 130-131.

See also:

ATKINSON, R.G. 1967. A modified soil percolator for zoospore production and infection in studies on zoosporic root pathogens. Can. J. Plant Sci. 47: 332-334.

Describes a useful device to study the effectiveness of zoospores as inoculum units.

VI. Measuring cankers caused by *Phytophthora* spp.

1. Remove outer bark to expose lesion.

2. Cover the entire lesion area, plus some surrounding healthy tissue (about 5 mm from lesion edges), with overlapping strips of transparent adhesive tape which can be written upon, e.g. Scotch brand Magic Transparent Tape #810.

3. Trace the entire outline on the tape with a pen or fine point marker.

4. Peel the tape from the lesion in one piece and transfer to a sheet or notebook. If the lesion is large, cut the tape in halves and measure separately.

5. Lesions area can be determined at a later date by following the traced outline with a polar-compensating planimeter, e.g. Salmoiraghi 236A (Lietz No. 3651-00), which give direct cm^2 measurements.

6. This method has been used to compare cankers caused by *Phytophthora* spp. on avocado, cacao, and other hosts.

KLURE, L. and G.A. ZENTMYER. 1976. Phytophthora Newsletter 4: 34.

Section 9:

Techniques for studying various

propagules of *Phytophthora* in soil

I. The Use of Fluorescent Brighteners in Soil Studies with
    Phytophthora

A great advancement in the ecological studies of soil-borne
microorganisms was made possible with the advent of fluorescent
brighteners. These vital stains are non-toxic, are readily absorbed by microorganisms, and are highly stable, showing no de-
tectable loss of fluorescence over long periods of time. These
stains can thus be used to stain specific microorganisms which
can later be recovered from the soil and conveniently viewed by
fluorescence microscopy.

1. Calcofluor

   The stilbene derivative "Calcofluor White M2R New"
   (disodium salt of 4, 4'-bis (4-anilino-6-diethylamino-
   s-triazin-2-ylamino) -2, 2'-stilbene-disulfonic acid),
   has been the main optical brightener used as a vital
   stain for microorganisms (Darken, 1962; Gisi & Schwinn,
   1976; Tsao, 1970). Calcofluor-labelled mycelium, sporangia, zoospores, and chlamydospores have fluoresced
   brightly when incorporated into soil and subsequently
   recovered for observation by fluorescence microscopy.
   Treatment with 100 ppm of calcofluor is sufficient to
   induce long-lasting fluorescence that can be transported
   to newly formed vegetative and reproductive structures.
   Aqueous solution of the brightener can be sterilized by
   millipore membrane filtration (0.22 μm or 0.45 μm pore
   size).

   The disadvantages of Calcofluor are its low pH, low
   solubility in water, and its propensity to stimulate
   sporangia production.

2. Diethanol

   Diethanol (Gisi, 1975), appears to be a good replacement
   for calcofluor due to its higher water solubility and
   more favorable pH. It can also be used at higher dosage
   rates than Calcofluor.

3. Acridine Orange

The fluorochrome, acridine orange, can be used to rapidly
scan soil smears for the presence of *Phytophthora*. A drop
of a 1% solution of this fluorochrome on a soil sample
will stain any mycelium and reproductive structures
present so that they can easily be viewed under fluorescence microscopy. This stain is particularly useful for
detecting pellucid bodies in oospores recovered directly
from soil.

References:

DARKEN, M.A. 1962. Appl. Microbiol. 10: 387-393.

GISI, U. 1975. Z. Pfl. Kr. Pfl. Schutz 82: 355-377

GISI, U. and F. SCHWINN. 1976. Microscopica Acta 77: 402-419.

TSAO, P.H. 1970. Soil Biol. Biochem. 2: 247-256.

II. Advantages and Disadvantages of some Fluorescent Brighteners

   A. Cellulose markers

   I. Calcofluor White M2R New   (American Cyanamid Co.)

      1. Has low water solubility.

      2. pH=10, making it unsuitable for most soil micro-
         biological studies.

      3. Inhibits mycelial growth above 300 ppm.

      4. Sporangia production of *P. cactorum* is stimulated (ap-
         prox. 33%) with Calcofluor at concentrations greater
         than 200 ppm.

   II. Diethanol

      1. Has high water solubility.

      2. pH=5.5.

3. Due to its hydroxyl groups, it has a high affinity for polymers such as cellulose.

4. Significantly stimulates mycelial growth between 100-500 ppm.

5. Rates above 1000 ppm reduce the growth rate of the mycelium.

6. Sporangia production of *P. cactorum* has been stimulated (76-105%) with diethanol at 1000 ppm.

Notes:

a) The papillum of the sporangium does not fluoresce with Calcofluor or diethanol, whereas septa, young hyphal tips, the pedicel, the plug of the sporangium, the chlamydospores, and the region of the wall where the germ tube emerges shows intensive blue fluorescence.

b) Chlamydospores and oospores change their color of fluorescence with increasing age from blue to yellow.

B. Plasma markers

I. Coumarin

Favors growth up to 100 ppm

II. Acridine Orange

Favors growth up to 10 ppm. It is toxic above 50 ppm.

In general, cellulose markers such as Calcofluor and diethanol are better suited for long range observations of test fungi in soil than dyes such as coumarin and acridine orange.

References:

ISI, U. 1975. Z. Pflanzenkrankheiten/Pflanzenschutz 82: 30-47, and 82: 55-377.

GISI, U. and F. SCHWINN. 1976. Microscopica Acta 77: 402-419.

III. Techniques for using Fluorescent Brightener

Calcofluor

1. Grow the fungus in a liquid medium containing the bright
   ener (100-300 ppm) or directly stain the fungal propagules before use.

2. The brightener is readily absorbed by the fungus and is
   non-toxic at the above concentrations.

3. Labeled propagules of *Phytophthora* introduced into soil
   can be recovered for observation using a soil-smear tech
   nique.

4. Under the fluorescence microscope, the labeled propagule
   and subsequently formed germ tube, mycelia, chlamydospores, sporangia and encysted zoospores all fluoresce
   with little interference from the soil mass or soil mi-
   croflora.

TSAO, P.H. 1970. Soil Biol. Biochem. 2: 247-256.

IV. Estimation of Disease Potentials of Citrus *Phytophthoras* i
    soil

1. Pulverise 500-800 ml of soil sample by hand to fine par
   ticles (2 mm or less) and mix thoroughly.

2. Place 50 ml of the soil in a "split cup" (Fig. 1A) and
   divide into 2 equal portions (Fig. 1B).

3. Put 1/2 into a paper cup and label 1/1.

4. Add 25 ml of an autoclaved soil (about pH 7.0) to the
   other 1/2 and mix thoroughly with a clean stick (Fig.
   1C) to provide a 1-in-2 dilution.

5. Repeat this procedure to give dilutions of 1/1, 1/2, 1/
   1/8, 1/16, 1/32, and 1/64 (Fig. 3).

6. Treat another 50 ml of the original soil in the same

manner to provide a replicate. Use sterile soil as controls.

7. Add 150 ml of deionized distilled water to each cup and place 2 lemon fruits (ripe or near-ripe stage into each cup).

8. Incubate at about $25^{\circ}C$ for 6 days.

9. One or more brown rot lesions on either or both of the two fruits in a cup give positive infection readings for that dilution cup.

10. The Disease Potential Index (DPI) of a given soil is defined as the reciprocal of the highest of the dilutions that yield brown lesions on the lemon fruit; e.g., positive readings at 1/32 dilution = DPI = 32.

Notes:

Test fruits should be picked from branches high enough to avoid field contamination of *Phytophthora* species from soil in the orchard. When test fruits are from an unknown origin, the fruits should be washed in O.1N HNO_3 solution to remove copper fungicide residue possibly on the surface. Fruits with surface injuries are unsatisfactory because of the possibility of infection by *Penicillium*, etc.

TSAO, P.H. 1960. Phytopathology 50: 717-724.

See also:

) STEINER, G.W. and R.D. WATSON. 1965. Use of surfactants in the soil dilution and plate count method. Phytopathology 55: 728-730.

) NASHAUM, C.J., G.B. LUCAS, and J.F. CHAPLIN. 1952. Phytopathology 42: 286 (abstr.).

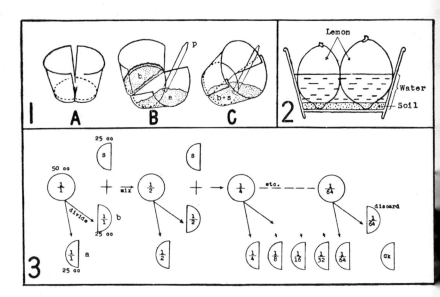

Fig. 4: Schematic diagram of the procedures involved in the dividing and t̲
mixing of the soil in the serial dilution end-point method.

Reproduced by permission from TSAO, 1960. Phytopathology 50: 717-724.

V.   Methods for Recovery of Propagules from Soil

     See the following publications for complete details.

1.   BRISTOW, P.R. and J.L. LOCKWOOD. 1975. J. Gen. Microbiol.
     90: 140-146. Describes a membrane filter method.

2.   GISI, U. 1975. Z. Pflanzenkrankh. Pflanzenschutz. 82: 355-
     377. Describes the use of fluorescent optical brighteners
     for direct observations.

3.   KO, W.H. 1971. Phytopathology 61: 437-438. Discusses the
     practicality of using a vertically illuminated microscope
     for direct observation of *Phytophthora* in soil.

4.   LINGAPPA, B.T. and J.L. LOCKWOOD. 1963. Phytopathology 53:
     529-531. Describes a plastic film method.

5. NASH, S.M., T. CHRISTOU, and W.C. SNYDER. 1961. Phytopathology 51: 308-312. Describes a soil smear method.

6. SNEH, B. 1977. Soil Biol. Biochem. 9: 65-66. Describes a nucleopore (polycarbonate) membrane filter system for observing the behavior of *Phytophthora* in soil.

7. TSAO, P.H. 1970. Soil Biol. Biochem. 2: 247-256. Describes the use of fiberglass squares and fluorescent brighteners for observation of soil *Phytophthora*.

VI. Method for Microscopic Observation of Living *Phytophthora* Propagules recovered from Soil

See, SNEH, B. 1977. Soil Biology and Biochemistry. 9: 65-66.

Section 10:

Disease Resistance

10A Screening for resistance

:. Evaluating *Phytophthora* root rot resistance in alfalfa
    varieties

\. Field Method

1. Select level area with relatively heavy soil.

2. If area is not naturally infected with *P. megasperma*,
    incorporate infested soil from infested field or green-
    house experiments into upper 15 cm.

3. Plant seed at about 75 viable seeds per 1.6 m row, in
    rows 30 cm apart (3 replications).

4. Make a stand count approximately 2 weeks after planting,
    while the seedlings are in the unifoliate stage.

5. When seedlings are about 4 weeks old, supplement rainfall
    with irrigation to keep the surface soil continuously at
    or near saturation for 2 to 3 weeks.

6. After the first 2-3 week wet period, allow the soil to
    dry.

7. Take notes and cultivate.

8. Re-wet the soil for an additonal 1 to 3 week period to
    reduce frequency of escapes, and also provide a more ac-
    curate evaluation of the degree of root rot on each plant
    due to the presence of larger roots.

9. When plants are dug up 4 months later, retain as much of
    the root system as possible. At least 30 cm of tap root
    should be retained.

10. Tie all plants from one row into a bundle.

11. Soak roots first in a tub of water and then wash with
    water under pressure.

12. Classify plants indivudually for degree of root rot.

13. Use the following rating scale:

    1 = roots are clean with no lesions and many small root-
        lets are present on the tap root.

    2 = only very small lesions (2 mm) are present on tap
        root, tap roots usually lack branch roots, and most

lesions occur at site where branch root had started growth.

3 = one or more large lesions on tap root, but none gird ling the tap root, the tips of a few larger branch roots are rotted off.

4 = extensive root lesions with the tap roots usually rotted off 15 cm or more below the crown.

5 = tap root almost completely destroyed but plant alive

6 = plants dead.

14. Plants rated 1 and 2 are considered resistant.

15. All tests should contain a resistant check and a susceptible check.

B. Greenhouse Method

1. Fill 20 cm deep water-tight containers with fine, steamed sand.

2. Infest the sand with fungus cultures grown on V-8 juice agar (200 ml V-8 juice, 20 g agar, 1 g $CaCO_3$, in 800 ml distilled water), in 9 cm petri dishes.

3. Blend 2 week-old cultures in water and add to the sand at a concentration of one petri dish to 500 cm^2 surface area. Mix the inoculum with the sand.

4. Plant seed directly into the sand.

5. Water the seedlings sparingly until they are well established (about 4 weeks).

6. Plug the drain holes in the tanks and add sufficient water daily to raise the water level to the sand surface

7. Maintain sand temperatures between 20-24°C since this i the optimum for *Phytophthora* root rot.

8. After 4 weeks of continual flooding, the plants can be evaluated for root rot as described in "Field Method" above.

9. Include resistant and susceptible check varieties for comparison.

Note:

Field tests give a more accurate evaluation of varieties because the roots are larger and easier to score, and more plants can be observed.

FROSHEISER, F.I. and D.K. BARNES. 1974. Standard tests to characterize pest resistance in alfalfa varieties. Agr. Res. Services. U.S.D.A. ARS-NC-19.

See also:

GRAY, F.A., R.B. HINE, M.H. SCHONHORST, and J.D. NAIK. 1973. A screening technique useful in selecting for resistance in alfalfa to *Phytophthora megasperma*. Phytopathology 63: 1185-1188.

An equally effective method for studying root rot of alfalfa is using the method developed by Tsao and Garber (see page 190). This essentially requires placing a clay saucer under each pot and flooding the saucer as well as the soil surface for three days, and for the next 4 days, inverting the saucer and watering from the top as required. This cycle is repeated for the duration of the experiment.

The degree of root rot on tap roots is then scored as follows: 0 = no disease; 1 = small lesions not more than 0.1 of the circumference of the tap root affected; 2 = slightly larger lesions encompassing not over 0.2 of the circumference of the root; 3 = lesions covering at least 0.2-0.5 of the circumference; 4 = lesions encompassing half to nearly all the root but plant still living; and 5 = plant wilted and entire root rotted.

IRWIN, D.C. 1966. Phytopathology 56: 653-657.

I. Screening for Resistance to *P. infestans*

. MAIN, C.E. and M.E. GALLEGLY. 1964. The disease cycle in relation to multigenic resistance of potato to late blight. Amer. Potato J. 41: 387-400.

A practical method of screening clonal selections in the greenhouse for mutligenic resistance to late blight is described. Visual disease-index measurements of disease development in the greenhouse is correlated with degree of multigenic resistance expressed by clones in the field in Mexico.

2. HODGSON, W.A. 1961. Laboratory testing of the potato for partial resistance to *Phytophthora infestans*. Amer. Potato J. 38: 259-264.

3. LAPWOOD, D.H. 1961. Laboratory assessments of the susceptibility of potato haulm to blight (*P. infestans*). European Potato J. 4: 117-127.

4. DEAHL, K.L., M.E. GALLEGLY, and R.J. YOUNG. 1974. Laboratory testing of potato tubers for multigenic resistance to late blight. Amer. Potato J. 51: 324-329.

These investigators have found that a correlation can be shown between tuber and foliar resistance of potato clones. This correlation was only noted when wounded tubers were inoculated with pathogenic races of *P. infestans*, 24-48 hrs after wound periderm formation at 20°C.

5. MALCOLMSON, J.F. 1976. Assessment of field resistance to blight (*Phytophthora infestans*) in potatoes. Trans. Brit. Mycol. Soc. 67: 321-326.

Methods for assessing field resistance to blight are described in detail. A description is also given of the scale 1 (susceptible) to 9 (resistant), used to indicate lesion size on leaves and other tissues. Note: In Black's system, 1 = resistant, and 5 = very susceptible.

III. Screening for glycoalkaloids in Potato

FITZPATRICK, T.J. and S.F. OSMAN. 1974. Method for determination of potato glycoalkaloids. Amer. Potato J. 51: 318-323.

Note:

Glycoalkaloids have been shown to be part of the resistance
mechanism in potatoes to pathogenic fungi. These compounds are
naturally present in potatoes and increase in concentration
under conditions of biological and environmental stress.

V. Interactions of Potato Blight with Potato Viruses

It has been reported that infection with potato viruses x, y,
a, and m, resulted in higher resistance of potato to blight incited by *P. infestans*.

With leafroll virus, on the other hand, reports indicate that a
more severe infection of potato by the blight organism (*P. infestans*), can occur; while other investigations indicate that
leafroll virus infection results in more resistance to *P. infestans* in the potato.

References:

MULLER, K.O. and J. MUNRO. 1951. Ann. Appl. Biol. 38: 765-773.

MIETKIEWICZ, J. 1974. Phytopathol. Z. 81: 364-372.

RICHARDSON, D.E. and D.A. DOLING. 1957. Nature 180: 866-867.

Disease Resistance

10B Genes for resistance

I. Alfalfa (*Medicago sativa* L.): *P. megasperma*

Alfalfa is a 4n species. Genetic control for resistance to *P. megasperma* is probably by one tetrasomic gene with incomplete dominance. Resistance is believed to be mainly controlled by the pm gene. Highly resistant plants have the nulliplex genotype (pm, pm, pm, pm), and highly susceptible plants are quadruplex (Pm, Pm, Pm, Pm). Simplex, duplex, and triplex genotypes are increasingly more susceptible.

FROSHEISER, F.I. and D.K. BARNES. 1973. Crop Sci. 13: 735-738.

LU, N. S.-J., D.K. BARNES, and F.I. FROSHEISER. 1973. Crop Sci. 13: 714-717.

II. Cacao (*Theobroma cacao*): *P. palmivora*

Extensive discussion on breeding for resistance to *P. palmivora* is given in "Phytophthora Disease of Cocoa," P.H. Gregory (ed.). Longman Group Ltd. New York and London. 1974. (see pages 219-268).

III. Cowpea (*Vigna sinensis* L. Endl. ex Hassk.): *P. vignae*

Resistance is inherited monogenically.

IV. Pepper (*Capsicum annum* L.): *P. capsici*

Resistance is governed by at least two distinct dominant genes that are inherited independently without additive effects.

POLACH, F.J. and R.K. WEBSTER. 1972. Phytopathology 62: 20-26.

SMITH, P.G. *et al*. 1967. Phytopathology 57: 377-379.

V. Potato (*Solanum tuberosum* L.): *P. infestans*

Resistance is controlled by single dominant genes (vertical
resistance), and by a combination of several lesser genes (horizontal or multigenic resistance). Over 15 pathogenic races are
known.

Potato varieties are allopolyploids (2n = 48), making inheritance a complex problem. Also, resistance of the foliage is not
always correlated with resistance of the tubers and presumably
different genes govern these characters. The wild species, *Solanum demissum* (2n = 72), has given high levels of resistance
to *S. tuberosum* in crosses between these two species. However,
the rapid appearance of new races of *P. infestans* has resulted
in the incorporation of horizontal or multigenic resistance into *S. tuberosum*. This type of resistance is evenly spread a-
gainst all races of the pathogen, but at a much lower level tha
that obtained by single dominant or vertically inherited genes.
Used in conjunction with vertical or dominant gene resistance,
horizontal or multigenic resistance can be very useful and long
lasting.

For a detailed discussion of pathogenic races and breeding for
resistance to *P. infestans,* see: J.E. Van der Plank (1968). Dis
ease Resistance in Plants. Academic Press. New York & London,
206 p.

VI. Safflower (*Carthamus tinctorius* L.): *P. drechsleri*

Resistance is controlled by a dominant gene, or a gene pair
(pH_2pH_2) in some varieties, and by a single recessive pair
(ph_1, ph_1) in other varieties. Recessive gene resistance par-
ricularly under extensive and prolonged irrigation, is superior
to the dominant resistance.

It appears that root resistance to *P. drechsleri* is inherited
through a single dominant gene, and hypocotyl resistance throug
a recessive gene. Pathogenic races are known.

DISEASE RESISTANCE 219

THOMAS, C.A. 1960. Phytopathology 50: 129-130.

THOMAS, C.A. 1970. Phytopathology 60: 63-64.

THOMAS, C.A. 1976. Plant Disease Reptr. 60: 123-125.

VII. Soybean (*Glycine max* [L.] Merr.): *P. megasperma* var. *sojae*

Resistance is controlled by a dominant gene Rps. There is evidence that other genes besides Rps influence resistance to the pathogen. The genetics now appear to be quite complex with new races (1-6), and at least three alleles being reported. Order of dominance is Rps > rps^2 > rps.

SCHMITTHENNER, A.F. 1972. Plant Disease Reptr. 56: 536-539.

VIII. Tobacco (*Nicotiana tobaccum* L.): *P. nicotianae* var. *parasitica*

Multigenic resistance as well as undesirable linkages are known. Monogenic dominant resistance has been transferred from two related species, *N. longiflora,* and *N. plumbaginifolia,* but this gene is susceptible to race 1 of the pathogen.

CLAYTON, E.E. 1969. Tobacco Sci. 168: 30-37.

Disease Resistance

10C Phytoalexins and antifungal compounds

Phytoalexins

The following antifungal compounds have been isolated from various host plant cells that had previously been challenged by an
incompatible race of *Phytophthora*.

Phytoalexin	Host	Pathogen	Authority
haseollin	French Bean	*P. infestans*	Jerome & Muller, 1958
ot named	Red Pepper	*P. infestans*	Muller, 1958
ishitin	Potato	*P. infestans*	Tomiyama *et al.*, 1968
ot named	*Peperomia* sp.	*P. nicotianae* var. *parasitica*	Siradhana *et al.*, 1969
yubimin	Potato	*P. infestans*	Ozeretskovskaya *et al.*, 1969
afynol	Safflower	*P. drechsleri*	Thomas & Allen, 1969
-hydroxybenzoic acid -hydroxybenzoic acid -hydroxy-3-methoxy-enzoic acid	Potato	*P. infestans*	Robertson, 1969
ot named	Strawberry Roots	*P. fragariae*	Mussell & Staples, 1971
-α-Hydroxyphaseollin	Soybean	*P. megasperma* var. *sojae*	Keen *et al.*, 1971
ytuberin	Potato	*P. infestans*	Varns & Kuc, 1971
apsidiol	Pepper fruit	*P. infestans*	Stoessl *et al.*, 1972
evitone	Cowpea	*P. vignae*	Partridge & Keen, 1976
ot named	Carnation	*P. nicotianae* var. *parasitica*	Ricci, 1977

ntifungal Substances: Borbonol

orbonol, an antifungal compound, has been isolated and char-
cterized from several species in the genus *Persea*, that show
igh resistance to *P. cinnamomi* (Zaki *et al.* 1976. Phytophthora
ewsletter 4: 27-28). Borbonal inhibits mycelial growth of *P.
innamomi* at a concentration of 1 ppm (3.2 μmole). It is be-
ieved that this substance may in part be responsible for re-
istance in some *Persea* spp. to *Phytophthora* root rot.

Section 11A:

Genetics of *Phytophthora*

I. Pathogenicity: The Race Concept

Black (1952) identified four dominant genes for resistance to
Phytophthora infestans in *Solanum demissum*. On the basis of his
genetic analyses he postulated 16 different races of *P. infestans* (0; 1.2; 3.4; 1.2.3; 1.2.3.4; etc.). Race 1.2.3.4 would
overcome resistance to any combination of the four dominant genes
in the host plant (Table 7). Since then, all 16 races have been
isolated in nature.

Toxopeus (1956) then postulated a gene-for-gene relationship for
the *P. infestans: Solanum* system. Several other dominant genes
have now been incorporated into the potato (*Solanum tuberosum*)
for resistance to potato blight. This has predictably resulted
in a concomitant increase in the number of new races of *P. infestans* isolated in nature.

Pathogenic races have also been identified in other *Phytophthora*
species e.g. *P. fragariae* (Strawberry red stele disease), and
. megasperma var. *sojae* (soybean root rot).

BLACK, W. 1952. Proc. Roy. Soc. Edinburgh 64(B): 312-352.

TOXOPEUS, H.J. 1956. Euphytica 5: 221-237.

II. Gene-for-Gene Relationships

It is postulated that for each gene determining resistance in
the host plant, there is a specific gene determining virulence
in the pathogen (Flor, 1956). In the ideal system, each host
locus controls either a resistant or a susceptible reaction to
the related locus in the pathogen, and each locus in the pathogen controls either a virulent or an avirulent reaction toward
the related locus in the host.

A gene-for-gene relationship has been described for the *Phytophthora: Solanum* system (see Table 7; for each new dominant gene
added to the potato, there is a new race of *P. infestans* that
develops to overcome the added resistance).

228   GENETICS

FLOR, H.H. 1956. Advances in Genetics 8: 29-54.

PERSON, C. 1959. Gene-for-gene relationships in host: parasite systems. Can
J. Botany 37: 1101-1130.

III. Types of Resistance

   1. Vertical Resistance (Monogenic, major gene, hypersensi-
      tivity).

      This type of single dominant gene resistance is effecti
      against certain races and ineffective against other rac
      of the pathogen. In potato, dominant gene resistance fo
      potato blight is soon overcome by new races of P. *in-
      festans*, partly due to the intense selective pressure
      exerted on the pathogen by the introduction of resistan
      plants. See page 218 for a discussion of vertical gene
      resistance in the breeding of potatoes for resistance t
      P. *infestans*.

   2. Horizontal Resistance (Polygenic, minor gene, field re-
      sistance, non-specific resistance, multigenic   resist-
      ance).

      This type of resistance reduces the amount of disease
      development by slowing the rate of increase of the dis-
      ease. The resistance is not race-specific (e.g. multi-
      genic resistance in potatoes against P. *infestans*). Ho
      zontal resistance is very useful when used in combinat
      with vertical resistance for effective control against
      many races of the pathogen (see page 218).

Table 7: The Interrelationship of Dominant Genes for Resistance to Late
Blight in the Potato and Pathogenic Races of *Phytophthora Infestans*
(the Gene-for-Gene Relationship)

Host Resistance genotype	Pathogenic races of *P. infestans*															
	0	1	2	3	4	1,2	1,3	1,4	2,3	2,4	3,4	1,2,3	1,2,4	1,3,4	2,3,4	1,2,3,4
	S	S	S	S	S	S	S	S	S	S	S	S	S	S	S	S
R_1	-	S	-	-	-	S	S	S	-	-	-	S	S	S	-	S
R_2	-	-	S	-	-	S	-	-	S	S	-	S	S	-	S	S
R_3	-	-	-	S	-	-	S	-	S	-	S	S	-	S	S	S
R_4	-	-	-	-	S	-	-	S	-	S	S	-	S	S	S	S
R_1R_2	-	-	-	-	-	S	-	-	-	-	-	S	S	-	-	S
R_1R_3	-	-	-	-	-	-	S	-	-	-	-	S	-	S	-	S
R_1R_4	-	-	-	-	-	-	-	S	-	-	-	-	S	S	-	S
R_2R_3	-	-	-	-	-	-	-	-	S	-	-	S	-	-	S	S
R_2R_4	-	-	-	-	-	-	-	-	-	S	-	-	S	-	S	S
R_3R_4	-	-	-	-	-	-	-	-	-	-	S	-	-	S	S	S
$R_1R_2R_3$	-	-	-	-	-	-	-	-	-	-	-	S	-	-	-	S
$R_1R_2R_4$	-	-	-	-	-	-	-	-	-	-	-	-	S	-	-	S
$R_1R_3R_4$	-	-	-	-	-	-	-	-	-	-	-	-	-	S	-	S
$R_2R_3R_4$	-	-	-	-	-	-	-	-	-	-	-	-	-	-	S	S
$R_1R_2R_3R_4$	-	-	-	-	-	-	-	-	-	-	-	-	-	-	-	S

S = Susceptibility; - = Resistance

Reproduced from M.E. GALLEGLY, Ann. Rev. Phytopathol. 6: 375-396.
By permission Annual Reviews Inc. Copyright 1968.

V. Origins of Asexual Variation

Variation for all morphological and physiological characters
have been documented for most *Phytophthora* species studied in
detail. The origins of these variations are the subject of much
speculation. In particular, the origin of new races of *P. infestans* in the absence of sexual reproduction is a puzzling one.

230

The frequency of new races would seem to preclude mutation as the obvious mechanism involved. The spontaneous 'mutation rate' reported by investigators is far in excess of that reported for other microorganisms. Also, the reversion rate (back-mutations) reported for isolates of *P. infestans* are greater than normally experienced by most microorganisms.

The following mechnisms have been postulated by various investigators to explain asexual variation and the aquisition of virulence genes in *Phytophthora*.

1. Mutation and selection
2. Somatic recombination (Parasexuality)
3. Heterocaryosis
4. Cytoplasmic control
5. Adaptive parasitism
6. Aggressiveness

1. Mutation

Abnormally high 'mutation' rates have been reported for virulence loci in *P. infestans*. The high rate of appearance of new races would postulate a unique mechanism(s) of mutation not ordinarily encountered in microorganisms. Investigators have both supported and discounted mutation as a primary mechanism for asexual variation in *Phytophthora*. Many investigators have reported a lack of success in inducing mutations in *Phytophthora* using a variety of physical and chemical methods. There is also evidence now available that *Phytophthora* is vegetatively diploid which would rule out the feasibility of a high mutation rate being responsible for asexual variation.

GALLEGLY, M.E. 1968. Annual Review Phytopathol. 6: 375-396.

2. Somatic recombination (Parasexuality)

Somatic recombination is reported to occur in low frequencies
thus requiring selective techniques to demonstrate its effects.
The frequency rate may be increased if the character observed
is controlled by many genes. Malcolmson (1970) isolated races
1.2.3.4.7, from sporangia of *P. infestans* produced on potato
leaf tissue previously inoculated with races 4 and 1.2.3.4; and
race 1.2.3.4.7.10 from inoculations with races 3.4.10 and 1.2.
4.7. Denward (1970) also demonstrated somatic recombination by
isolating race 1.2.3.4. of *P. infestans* from potato leaves previously inoculated with races 1.3. and 1.4. Similar results had
been previously reported by Leach and Rich (1969). The sexual
stage of *P. infestans* is not known to occur in Europe, thus
ruling out sexual recombination as one of the factors involved.

DENWARD, T. 1970. Hereditas 66: 35-48

LEACH, S.S. and A.E. RICH. 1969. Phytopathology 59: 1360-1365.

MALCOLMSON, J.F. 1970. Nature 225: 971-972.

3. Heterocaryosis

The possibility of heterocaryon formation would seem to be
feasible from the reported observations of anastomosis (union
of hyphae from different mycelium) in *Phytophthora*, by Stephenson *et al.* (1974). Their observations tended to support postula-
tions of heterocaryosis in *Phytophthora* by several investigators.
The first direct confirmatory evidence of heterocaryosis was
presented by Long and Keen (1977). They demonstrated heterocaryosis in *P. megasperma* var. *sojae* by the synthesis and de-
gradation of various auxotrophic and drug resistant mutants.
However, they reported that the frequency of heterocaryosis did
not appear to be high enough to be of importance in nature unless other mechanisms such as polyploidy, somatic crossing over,
and/or mutation were also operative.

LONG, M. and N.T. KEEN. 1977. Phytopathology 67: 670-674.

STEPHENSON *et al.* 1974. Phytopathology 64: 149-150.

4. Cytoplasmic control

Cytoplasmic control for variation in growth rate, sporulation,
and pigmentation have been shown for such fungi as *Aspergillus*
and *Penicillium*. Much less is known about cytoplasmic control i
Phytophthora due to the complex nature of the problem. Caten
and Jinks (1968) found extensive variation in growth rate and
sporangium production among single zoospore, single sporangium
and single hyphal tip subcultures of *P. infestans*. They conclu
ed that cytoplasmic differences in the original mass isolates
appeared to be the most likely cause of the observed variation.
A similar conclusion was reached by Caten (1971) in trying to
explain single zoospore variation and attenuation of strains i
culture, and by MacIntyre and Elliott (1974) in their studies
of the variation in growth-rate in asexual and sexual propaga-
tion of *P. cactorum*.

CATEN, C.E. 1971. Trans. Brit. Mycol. Soc. 56: 1-7.

CATEN, C.E. and J.L. JINKS. 1968. Can. J. Botany 46: 330-348.

MACINTYRE, D. and C.G. ELLIOTT. 1974. Genet. Res. 24: 295-309.

5. Adaptive parasitism: 'Plasticity' of *Phytophthora*

The apparent 'plasticity' of *P. infestans* isolates was first d
scribed by Reddick and Mills (1938). They found that the virulence of *P. infestans* could be increased by serial passage
through plants of a resistant variety. Mills (1940) also increased the virulence of *P. infestans* isolates from potato by
7 serial passages through tomato foliage. Other workers (de Br
1951 and Black 1952), confirmed these results. This 'adaptabil
ty' or 'plasticity' is not confined to *P. infestans*. Increased
virulence has also been reported with isolates of *P. nicotiana*
var. *parasitica* after 8 serial passages through resistant and
susceptible tobacco plants (Dukes and Apple, 1961). There is

however, evidence of failure by several investigators to effect
changes in specific pathogenicity by repeated host passage. The
changes reported above are believed by many workers to be attributable to mutation, since new races obtained by host passage
were highly stable.

It should be noted that with *P. infestans*, the changes in specific pathogenicity by host passage has been successful only
when juvenile, senescent, or abnormal foliage was used. Plant
tissues carrying dominant R gene, as well as multigenic resistance were less likely to effect changes in the fungus by serial
host passage.

DUKES, P.D. and J.L. APPLE. 1961. Plant Disease Reptr. 45: 362-365.

GALLEGLY, M.E. 1968. Annual Review Phytopathol. 6: 375-396.

MILLS, W.R. 1940. Phytopathology 30: 17.

REDDICK, D. and W. MILLS. 1938. Amer. Potato J. 15: 29-34.

. Aggressiveness

This component of pathogenicity expresses the virulence differences often observed between the same races of the pathogen, e.
. *P. infestans*. Aggressiveness characterizes not only the degree of pathogenicity on an individual plant, but also the potential of an isolate in epiphytotics.

Prolonged maintenance of the pathogen in artificial culture
often leads to a loss in aggressivity. This loss in aggressivity can be restored to its original level by serial transfer of
the fungus on the leaves or tubers (in the case of *P. infestans*),
of the host variety from which it was originally isolated,
Jinks and Grindle, 1963, Heredity 18: 245-264).

See Van der Plank (1968), Disease Resistance in Plants, Academic Press. N.Y. for a discussion of aggressivity in relation to
the virulence of *P. infestans* isolates.

7. Anastomosis

Several of the above methods for asexual variation would be fea
sible if anastomosis (fusion between branches of the same or
different hyphae), occurred in *Phytophthora*. The speculation re
garding anastomosis in *Phytophthora* was finally laid to rest
when Stephenson *et al.* (1974), presented evidence for anastomosis in cultures of *P. capsici*, *P. drechsleri* and *P. infestans.*
This work confirmed and extended the short report on anastomos.
in *P. infestans* by Wilde (1961). The occurrence of anastomosis
thus raised the possibility that parasexuality, adaptive parasitism, and heterocaryosis, could all be operative in *Phytoph-*
thora.

STEPHENSON, L.W., D.C. ERWIN, and J.V. LEARY. 1974. Phytopathology 64: 149-
150.

WILDE, P. 1961. Archiv. Mikrobiol. 40: 163-195.

Notes:

It is now known that:

1. Depending on environmental conditions and the selectiv
   pressures exerted by the host, any one, or a combinatio
   of the above mechanisms can be operative in the expres-
   sion of asexual variation in *Phytophthora*.

2. Virulence genes may be lost in *Phytophthora* by mutation
   or dissociation of heterocaryons.

V.   Inheritance of Compatibility or Mating Types.

The inheritance of compatibility types in *Phytophthora* has lon
been a vexing problem. Ratios obtained in several genetic
studies using compatibility types A^1 and A^2 as markers have
failed to give a clear indication of the mechanism of inheritance of these characters. Data obtained have been interpreted
as indicating cytoplasmic or nuclear control of mating type.

These conflicting interpretations stem from the major problems
of most genetic studies undertaken with *Phytophthora*, i.e. extremely low germination of oospores in succeeding generations,
and the difficulty of establishing sufficient viable progeny
from germinating oospores to permit genetic analyses.

Galindo and Gallegly (1960), reported that compatibility types
A^1 and A^2 of *P. infestans* existed in a 1:1 ratio in Mexico,
suggesting that the two factors are allelic. Near 1:1 ratios
have also been reported for compatibility types of *P. capsici*,
P. meadii, and *P. nicotianae* var. *parasitica* (Polach and Webster,
1972; Savage *et al.*, 1968). However, Gallegly (1968), reviewed
several reports showing a skewed relationship for mating types,
favoring a preponderance of the A^1 type.

It now appears that mating types are controlled by a more complex mechanism than a single pair of alleles (Romero, 1969), or
by more than one closely linked locus inherited as a single fac-
tor (Polach and Webster, 1972). Timmer *et al.* (1970), on the
other hand, postulated an even more complex system for the
determination of mating types. Their system involved three independently segregating pairs of alleles, Aa, Bb, and Cc, with
bc assumed to be lethal. A combination of these genotypes in
crosses with A^1 and A^2 types would give a number of different
genotypes which would explain the skewed ratios reported by
several investigators.

Cytoplasmic control has also been postulated (Gallegly, 1968).

GALINDO, A.J. and M.E. GALLEGLY. 1960. Phytopathology 50: 123-128.

GALLEGLY, M.E. 1968. Annual Review Phytopathol. 6: 375-396.

POLACH, F.J. and R.K. WEBSTER. 1972. Phytopathology 62: 20-26.

SAVAGE *et al.* 1968. Phytopathology 58: 1004-1021.

TIMMER *et al.* 1970. Amer. J. Botany. 57: 1211-1218

VI. Interspecific Hybridization

All heterothallic *Phytophthora* species freely inter-cross with
each other producing mature oospores when mated on a suitable
medium (Table 8). However, oospores from such crosses germinate
in very low frequencies making genetic analyses of the progeny
difficult. Boccas and Zentmyer (1976) reported successful germination (5%) of oospores produced between a cross of *P. nicotianae* var. *parasitica* and *P. cinnamomi*. Evidence of recombination was obtained in relation to mating type, pathogenicity to
citrus and *Persea indica*, and response to temperature.

BOCCAS, B. and G.A. ZENTMYER. 1976. Phytopathology 66: 477-484.

SAVAGE *et al.* 1968. Phytopathology 58: 1004-1021.

VII. Stabilizing Selection

Van der Plank (1968), put forward the idea that complex races
of a pathogen contain unnecessary genes for virulence and are
thus less fit to survive than simple races, e.g. race 1.2.3.4.
of *P. infestans* is less able to survive compared to simple race
such as 4, or 1.2.

A complete discussion on stabilizing selection with regard to
races of *P. infestans* and vertical resistance in *Solanum tuberosum* is given in Van der Plank (1968), Disease Resistance in
Plants, Academic Press, N.Y. See pages 59-63.

VIII. Ploidy of *Phytophthora*

There is currently considerable controversy regarding the basi
mechanisms of inheritance in *Phytophthora*. The vegetative mycelial stage has been interpreted as being haploid, diploid,
and even polyploid. Published research offers evidence of sexu
recombination, heterocaryosis, vegetative recombination, and
cytoplasmic inheritance in various *Phytophthora* species.

Table 8: Examples of interspecific hybridization between heterothallic *Phytophthora* species

Compatibility Type A^2

Compatibility Type A^1	P. arecae	P. cambi-vora	P. cap-sici	P. cinna-nomi	P. citroph-thora	P. colo-casiae	P. cryp-togea	P. drech-sleri	P. infes-tans	P. meadii	P. palmi-vora	P. nicot. var. para.
P. cambivora	+	+	+	+	+[a]	+	+	+	+[a]	+	+	+
P. capsici	+	+	+	+[a]	+	+	+	+	+[a]	+	+	+
P. cinnamomi	+	+	+	+	+	+	+	+	+[a]	+	+	+
P. citrophthora	+	-	+	+	+	+	-	+	+[a]	+	+	+
P. cryptogea	+	+	+	+	+	+	+	+	+	+	+	+
P. drechsleri	+	+	+	+	+	+	+	+	+	+	+	+
P. infestans	+	+	+	+[a]	-	+[a]	-	+	+	+[a]	+[a]	+
P. meadii	+	+	+	+	+	+	+	+	+[a]	+	+	+
P. palmivora	+	+	+	+	+	+	+	+	+[a]	+	+	+
P. nicotianae var. parasitica	+	+	+	+	+	+	+	+	+	+	+	+

+ = oospores formed. − = oospores not formed.

a = gametangia often abortive, i.e. no oospores are formed.

Modified from SAVAGE *et al.*, Phytopathology 58(1968), by permission.

Genetic evidence for haploidy has been reported by Gallegly
(1968), Laviola (1968), and Romero and Erwin (1969), with *P.
infestans* as test organism and genetic markers comprising of
compatibility types and pathogenicity race designations. Segregation of the pathogenicity markers generally fitting a 1:1 ratio in the F_1 generation led to the tentative conclusion of
vegetative haploidy in *P. infestans*. Timmer (1969) studying the
inheritance of auxotrophic mutants of *P. capsici* supported the
haploid hypothesis. However, it was necessary to postulate the
existence of suppressors to explain the data on the basis of
vegetative haploidy. Satour and Butler (1968) reported 1:1 F_1
segregation for mating type, pathogenicity, and morphological
markers in *P. capsici*, while Polach and Webster (1972) obtained
1:1 segregation ratios in the F_1 for many but not all of the
pathogenicity markers used.

Genetic evidence supporting vegetative diploidy has been given
by Shaw and Khaki (1971) for *P. drechsleri*, using drug resistant
mutants as markers. They obtained no segregation in the F_1 generation, and segregation in the F_2 fitting a 3:1 ratio; a 1:1
ratio in a back-cross to the wild-type parent; and no segregation in back-cross to the drug resistant parent. Elliott and
MacIntyre (1973) also produced evidence for diploidy using *P.
cactorum* as a test organism and auxotrophic mutants as genetic
markers. Boccas (1973), also supported the diploid hypothesis
using *P. syringae* and morphological traits. Long *et al.* (1977)
in a very thorough study using large numbers of progeny, obtained data that indicated that *P. megasperma* var. *sojae* is
diploid in the vegetative state. Four classes of recombinant
progeny, as would be expected from a vegetatively diploid organism, were obtained when first generation oospores were selfed.

There is also conflicting cytological evidence regarding the
position of meiosis in the life cycle of *Phytophthora* species
(see Cytology section and Table 9).

Conclusions

The preponderance of the evidence indicates that *Phytophthora* species are unique in being vegetatively diploid.

The reasons for the conflicting interpretations of data can be
summarized as follows:

Genetical data

1. Poor germination of oospores, making valid analyses difficult. Analyses have been made with progeny resulting
   from oospore germination as low as 0.5%.

2. Too few genetic markers and the uncertainty of whether
   markers such as mating type and pathogenicity race designations are nuclear rather than cytoplasmic.

3. The failure to make reciprocal crosses and to replicate
   crosses.

4. The use of many different species and isolates of *Phytophthora* by various investigators, and the lack of data
   confirmation by independent workers.

5. The difficulty in obtaining mutants by conventional muta-
   genesis techniques.

6. The possibility of selfed oospores skewing ratios in
   matings between heterothallic species.

7. The utilization of wild, native *Phytophthora* for genetic
   work.

Useful references:

DAY, P.R. 1973. Genetics of Host-Parasite Interaction. W.H. Freeman & Co.,
San Francisco, Ca. U.S.A. 238 p.

DICK, M.W. and WIN-TIN. 1973. The development of cytological theory in the
Oomycetes. Biol. Rev. 48: 133-158.

GALLEGLY, M.E. 1968. Genetics of Pathogenicity of *Phytophthora infestans*.
Ann. Rev. Phytopathol. 6: 376-396.

240

Table 9: Ploidy of *Phytophthora*

Method	Species	Ploidy	Authority	Journal
Cytological	*cactorum*	diploid	Buddenhagen ('58)	Am. J. Botany
Cytological	*erythrosep. cactorum*	diploid	Sansome ('65)	Cytologia
Cytological	*drechsleri*	diploid	Galindo & Zentmyer ('67)	Phytopathol.
Genetical	*cactorum*	diploid	Shaw & Elliott ('68)	J. Gen. Micro.
Genetical	*capsici*	diploid	Satour & Butler ('68)	Phytopathol.
Genetical	*infestans*	haploid	Laviola ('69)	Diss. Abstrs.
Genetical	*infestans*	haploid	Romero & Erwin ('69)	Phytopathol.
Cytological	*parasitica*	diploid	Huguenin & Boccas ('70)	Compt. Rendu.
Genetical	*capsici*	haploid	Timmer *et al.* ('70)	Am. J. Botany
Genetical	*drechsleri*	diploid	Shaw & Khaki ('71)	Genet. Res.
Genetical	*syringae*	diploid	Boccas ('72)	Compt. Rendu.
Theoretical analyses	*Phytoph.* spp.	diploid	Dick & Whitehouse ('72)	European disc. group. Bari.
Lit. Review	*Phytoph.* spp.	diploid	Dick & Win-Tin ('73)	Biol. Review
Genetical	*cactorum*	diploid	Elliott & MacIntyre('73)	T. Br. My. Soc.
Genetical	*para.x cinna.*	diploid	Boccas ('73)	Fruits
Cytological	*infestans*	haploid	Laviola ('74)	Ann. Agric. Bar
Cytological	*infestans*	haploid	Laviola & Portacci ('74)	Phytopath. Medi
Cytological	*meg.v.sojae*	haploid	Stephenson *et al.* ('74)	Can. J. Botany
Cytological	*capsici*	haploid	Stephenson *et al.* ('74)	Can. J. Botany
Cytological	*megasperma*	diploid	Sansome ('74)	T. Br. My. Soc.
Cytological	*cinamomi infestans drechsleri*	diploid	Brasier & Sansome ('75)	T. Br. My. Soc.
Cytofluoro-metric	*drechsleri*	diploid	Mortimer & Shaw ('75)	Genet. Res.
Cytological	*capsici*	diploid	Sansome ('76)	Can. J. Botany
Cytological	*capsici*	diploid	Maia *et al.* ('76)	Annales Phyt.
Kill curves	*palmivora parasitica*	diploid	Pellegrin ('76)	Phytoph. Newslt
Genetical	*meg.v.sojae*	diploid	Long & Keen ('77)	Phytopathol.
Cytological	*infestans*	polyploid	Sansome ('77)	J. Gen. Micro.

11B: Genetic techniques

I. Mutagenic agents used with *Phytophthora*

   1. N'-nitro-N-nitrosoguanidine (NNG)

      A. 1. Shake zoospores suspension for 90 sec. to induce loss of motility.

          2. Add to flasks containing tris (hydroxymethyl) aminoethane (tris) buffer at a concentration of 0.2 $\underline{M}$, pH 10.

          3. After 30 min., add the buffered suspension to flasks containing crystals of NNG at five concentrations from 0.5-1.0 mg/ml.

          4. Shake gently for 5 hrs.

          5. Filter suspension through Whatman No. 4 filter paper and plate on pea extract medium.

ELLIOT, C.G. and D. MACINTYRE. 1973. Trans. Br. Mycol. Soc. 60: 311-316.

      B. 1. Produce zoospores and separate from hyphal fragments by passing through filter paper.

          2. Encyst the zoospores by agitation and concentrate by centrifugation.

          3. To 90 ml of cyst suspension in H_2O in a conical centrifuge tube, add 1 ml of 300 μg/ml NNG.

          4. Agitate for 10 min.

          5. Centrifuge for 3 min., decant NNG, dilute and wash by centrifugation.

          6. Screen cysts for mutants (see below).

LONG, MARGARET. 1975. Department of Plant Pathology. University of California, Riverside.

      C. Ultra-violet radiation (UV).

          Time required to kill 95-99% of the zoospores varies with type of UV lamp and distance from lamp to petri dish. Irradiating zoospores prior to encystment is more likely to induce mutants.

See: CLARKE, D.D. and N.F. ROBERTSON. 1966. European Potato J. 9: 208-215.

II. Screening for auxotrophic mutants

    A. Filtration Enrichment

        1. Inoculate liquid minimal medium with NNG treated cysts

        2. Allow prototrophs to grow and filter by centrifugation

        3. Overlay complete medium plates and screen colonies for auxotrophy.

        4. Transfer plug to minimal medium.

        Comments:

        Requires large initial numbers of zoospores.

    B. Plate wash technique

        1. Overlay 1 ml of NNG-treated cysts on CV8A.

        2. 12 hrs later, add 5 ml H_2O to wash surface of plate.

        3. Add this wash to next plate, and continue procedure for all plates.

        4. After the last plate is washed, overlay the wash water on CV8A and screen resultant colonies.

        Comments:

        Prototrophs germinate and establish, thus remaining attached to the surface of the plate. The non-germinators are removed in the wash water.

LONG, MARGARET. 1975. Department of Plant Pathology, University of California, Riverside.

See also, ANDREOLI, C. and J. PONCHET. 1974. Annales Phytopathol. (Paris) 6: 255-264.

III. Sorbose for Facilitating Viable Counts of Zoospores

1. Add 0.5% W/V L-sorbose to the medium, before plating out
   zoospores.

2. L-sorbose restricts linear growth, but does not affect
   viability.

Comment:

Without L-sorbose, zoospores plated on nutrient agar germinate to produce dense colonies of profusely branched hyphae
or diffuse colonies with sparsely branched hyphae, making
counts of % viable zoospores difficult.

KHAKI, I.A. 1974. Phytophthora Newsletter. 2: 17-18.

Section 12:

Cytology and electron microscopy

A. Cytology

Cytological studies of the genus *Phytophthora* have been the subject of much controversy. Interpretations of chromosomal configurations in somatic and gametangial nuclei have led various
investigators to conclude that the fungus is either haploid,
diploid, or polyploid, in the vegetative stage (Table 9). Even
studies using the same isolate of *P. capsici* have resulted in
dramatically different conclusions as to the ploidy of this
fungus; Stephenson *et al.* (Can. J. Bot. 52: 2141-2143, 1974),
concluded from their extensive cytological investigations that
P. capsici is haploid in the vegetative stage, while Sansome's
Can. J. Bot. 54: 1535-1545, 1976) cytological studies indicate
that *P. capsici* is vegetatively diploid. Other interpretations
of nuclear events in *Phytophthora* includes: 1) the idea that
true vegetative amitosis occurs with subsequent reconstruction
of the genome during gametangial production; 2) dispersion of
DNA from the vegetative nucleus followed by a reconstruction of
the vegetative nucleus; 3) linear alignment of chromosomes par-
llel to the long axis of the spindle. The spindle itself is
postulated to be parallel to the longitudinal axis of the hypha
necessitated by the limitations of space within the hypha; and
4) the chromosomes are believed to be linked in chains in the
somatic mycelium and in the gametangia. These difficulties in
interpretation have arisen largely due to: 1) the extremely
small size of the chromosomes making observations an extremely
difficult task in many instances; 2) the use of stains which
have not sufficiently differentiated the nuclear material; 3)
the difficulty of being able to follow nuclear events throughout the life cycle of *Phytophthora* due to a) the difficulty of
fixing and staining the thick-walled oospores, both for light
and electron microscopic studies and b) the problem of obtaining sufficient numbers of germinating oospores for cytological
studies; 4) trying to interpret *Phytophthora* cytology within
the framework of classical mitosis and meiosis to the exclusion
of new or novel systems which might offer a more plausible ex-
planation of observed cytological events.

I. The Sporangium

The sporangium is believed to be mutlinucleate as a result of
nuclear migration and subsequent mitotic divisions. Although
nuclear degeneration is known to occur in the sporangium, the
occurrence of heterocaryosis and hyphal anastomosis make single
sporangia an unsafe technique for the establishment of pure
genetic cultures. Hemmes and Hohl (Can. J. Bot. 9: 1673-1675,
1973), believed that the observed nuclear degeneration might be
a method for providing the remaining nuclei with specific pre-
cursors, and/or, that this nuclear migration might be a method
of maintaining a constant nuclear-cytoplasmic ratio.

II. The Gametangia

Evidence is now overwhelming that meiosis occurs in the oogoni
and in the antheridium. The life cycle of *Phytophthora* can
therefore be considered to be haplobiontic type B (Dick and
Win-Tin, 1973. Biol. Rev. 48: 133-158). The gametangia are multinucleate as a result of nuclear migration from the mycelium
into the oogonium and antheridium. Nuclear degeneration occurs
following meiosis in the gametangia. Excellent electron micro-
graphs of this phenomenon have been published by Hemmes and
Bartnicki-Garcia (Arch. Microbiol. 103: 91-112, 1975). A single
nucleus is then introduced from the antheridium into the oogon
um by means of a fertilization tube.

III. The Oospore

The mature oospore has a large refractive body and two pelluci
spots within which are contained the two gametic nuclei. These
two gametic nuclei are believed to fuse prior to germination.

A diagrammatic representation of the nuclear cycle in *Phytophthora* is shown in Fig. 6.

Useful References

.. DICK, M.W. 1969. Morphology and taxonomy of the Oomycetes, with special
reference to Saprolegniaceae, Leptomitaceae, and Pythiaceae. I. Sexual
reproduction. New Phytol. 68: 751-775.

. DICK, M.W. 1972. Morphology and taxonomy of the Oomycetes, with special
reference to Saprolegniaceae, Leptomitaceae, and Pythiaceae. II. Cyto-
genetic systems. New Phytol. 71: 1151-1159.

. DICK, M.W. 1973. The development of cytological theory in the Oomycetes.
Biol. Rev. 48: 133-158.

. WIN-TIN and M.W. DICK. 1975. Cytology of Oomycetes. Archiv. Microbiol.
105: 283-293.

Cytology of Individual Species

1. *P. cactorum*

SANSOME, E. 1966. Meiosis in the sex organs of the
Oomycetes, *in* Chromosomes Today, Vol. I, (C.D. Darling-
ton and K.R. Lewis, eds.). Oliver & Boyd. Edinburgh.

2. *P. capsici*

MAÏA, N., P. VENARD, and F. LAVRUT. 1976. Etude des divisions mitotiques et méiotiques du cycle de *Phytophthora
capsici*. Annales Phytopath. 8: 141-146.

SANSOME, E. 1976. Gametangial meiosis in *Phytophthora
capsici*. Can. J. Botany 54: 1535-1545.

STEPHENSON, L.W., D.C. ERWIN, and J.V. LEARY. 1974.
Meiotic configurations in the oospores of *Phytophthora
capsici*. Can. J. Botany 52: 2141-2143.

STEPHENSON, L.W., D.C. ERWIN, and J.V. LEARY. 1974. Cytology of somatic and gametangial nuclei in *Phytophthora
capsici* and *P. megasperma* var. *sojae*. Can. J. Botany 52:
2055-2060.

3. *P. cinnamomi*

BRASIER, C.M. and E. SANSOME. 1975. Diploidy and gametangial meiosis in *Phytophthora cinnamomi, P. infestans,*
and *P. drechsleri*. Trans. Brit. Mycol. Soc. 65: 49-65.

4. *P. drechsleri*

GALINDO, A.J. and G.A. ZENTMYER. 1967. Genetics and
cytology of *Phytophthora*. Nature, Lond. 214: 1356.

GALINDO, A.J. and G.A. ZENTMYER. 1967. Genetical and
cytological studies of *Phytophthora* strains pathogenic
to pepper plants. Phytopathology 57: 1300-1304.

See also: BRASIER and SANSOME (1975), ref. under *P. cinnamomi*.

5. *P. erythroseptica*

MARKS , P.A. 1918. The morphology and cytology of the
sexual organs of *Phytophthora erythroseptica*. Pethyb.
Ann. Bot. 32: 115-153.

See also: SANSOME (1966), ref. under *P. cactorum*.

6. *P. infestans*

MARKS, G.E. 1965. The cytology of *Phytophthora infestan*
Chromosoma 16: 681-692.

LAVIOLA, C. 1974. On the fertilization tube of *Phytophthora infestans* (Mont.) de Bary. Phytopath. Meditteran.
13: 55-58.

LAVIOLA, C. 1975. On the origin of multinuclearity of
sporangia in *Phytophthora infestans* (Mont.) de Bary.
Phytopath. Mediterran. 14: 30-31.

LAVIOLA, C. and M. PORTACCI. 1974. Osservazioni citolo-
giche sulle strutture sessuali di *Phytophthora infestan*
(Mont.) de Bary. Phytopath. Mediterran. 13: 87-92.

SANSOME, E. 1977. Polyploidy and induced gametangial fo
mation in British isolates of *Phytophthora infestans*. J
Gen. Microbiol. 99: 311-316.

SANSOME, E. and C.M. BRASIER. 1973. Diploidy and chromo

somal structural hybridity in *Phytophthora infestans*. Nature 241: 344-345.

SIVAK, B. 1973. The nature of the 'nuclear cap' and nuclear behavior in *Phytophthora infestans*. Can. J. Botany 51: 1983-1984.

7. *P. megasperma*

SANSOME, E. and C.M. BRASIER. 1974. Polyploidy associated with varietal differentiation in the *megasperma* complex of *Phytophthora*. Trans. Brit. Mycol. Soc. 63: 461-467.

8. *P. nicotianae* var. *parasitica*

HUGUENIN, B. and B. BOCCAS. 1970. Etude de la caryocinese chez le *Phytophthora parasitica* Dastur. C.R. Acad. Sci. Paris 271: 660-663.

9. *P. palmivora*

SANSOME, E., C.M. BRASIER, and M.J. GRIFFIN. 1975. Chromosome size difference in *Phytophthora palmivora*, a pathogen of cocoa. Nature 255: 704-705.

SANSOME, E. 1977. Report on the cocoa *Phytophthora* workshop. Rothamsted Expt. Sta. England, PANS. 23: 107-110.

General references

SANSOME, E.R. 1965. Meiosis in diploid and polyploid sex organs of *Phytophthora* and *Achlya*. Cytologia 30: 103-117.

WIN-TIN. 1972. Cytology and cytotaxonomy of Oomycetes. Ph.D. Thesis, University of Reading, England.

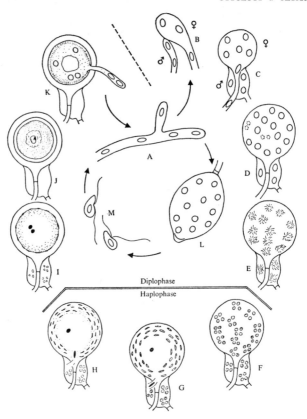

Fig. 6: A diagrammatic representation of the nuclear cycle in *Phytophthora*, as seen in *P. infestans*. Not to scale. A, Somatic mycelium with diploid nuclei (A1 and A2 mating type in a heterothallic species). B, Formation and association of antheridial (♂) and oogonial (♀) incepts. C, Emergence of the young oogonium (following penetration of the amphigynous antheridium by the oogonial incept); migration of nuclei into the developing oogonium. D, Abortion of nuclei in pairs in the expanded oogonium (c. 38μm diam): increase in size and remaining nuclei. E-F, Meiosis in the oogonium and antheridium; near-synchronous first division with a long prophase followed by a very short second division with no increase in nuclear size between the two divisions resulting in formation of tetrads of haploid nuclei. G, Beginning of oosphere delimitation; migration of the male gametic nucleus into the oogonium from the antheridium. H, Penetration of the oosphere by the male nucleus. I, Association of the male and female gametic nuclei in the oosphere. J, Formation of the thick walled oospore (c. 30μm diam) with a diploid zygotic nucleus : the 'pellucid spot'. K, Germination of the oospore to give the diploid vegetative stage. L, Diploid multinucleate sporangium (c. 29x19μm). M, Diploid uninucleate zoospores.

C. Cytological Techniques

I. Staining gametangia

1. Keep samples of spores to be fixed in iced water for
   1-2 hours, to ensure turgidity and to arrest divisions
   at metaphase.

2. Transfer some samples to water at room temperature for
   10-15 min. after cold treatment to obtain anaphases.

3. Blot off excess water after ice water treatment and
   fix for 30-45 min. in 3:1 ethanol:acetic acid.

4. Transfer to a mixture of 1 part ethanol to 1 part
   ether to remove oily substances.

5. Remove ether with acetic acid-alcohol or absolute
   alcohol.

6. Store material in 70% alcohol or water.

7. Select the material for staining by microscopic examination.

8. Before staining, immerse agar pieces in 1% citric acid
   for 10 min. This results in a swelling of the oogonium
   walls and greatly facilitates staining.

9. Rinse out the citric acid and transfer to 45-60%
   acetic acid for staining.

STAINS

A. Acetocarmine

1. Transfer agar pieces containing the spores from 45%
   acetic acid to a covered dish containing acetocarmine
   solution, to which a drop of ferric acetate solution
   is added.

2. Leave in the stain for 20-40 min.

3. Mount in a drop of fresh acetocarmine.

B. Acetoorcein

1. Make a saturated solution of acetoorcein by adding
   excess powdered orcein to a hot 60% acetic acid solution.

2. After cooling and filtering, make a weaker stain from
   the undissolved stain.

3. Transfer agar pieces containing the gametangia to a
   mixture of the weak and strong orcein in a covered
   dish for 3-4 hours.

4. Mount samples in a drop of orcein: melt agar by careful warming of the slide.

C. Lu's haematoxylin squash method

1. Hydrolyze material in 5N HCl for 10 min.

2. Transfer to the mordant, 0.5% iron alum in 50% propionic acid for 10 min.

3. Rinse in 50% propionic acid and transfer to stain.

4. Stain in 2% haemotoxylin and 50% propionic acid for
   2-5 min.

5. Carry out differentiation in 45% acetic acid containing a few drops of ferric acetate.

6. Mount in 45% acetic acid.

SANSOME, E. 1976. Can. J. Botany. 54: 1535-1545.

II. Staining *Phytophthora* nuclei with fluorochrome BAO

The following technique has been used successfully for cytological studies of nuclei of *P. drechsleri*.

1. Fix material in Newcomer's fixative, overnight. (Newcomer, E.H. 1953. Science 118: 161).

2. Take to water through the alcohol series.

3. Hydrolize in either 1N HCl at 60°C, or 6N HCl at 20°C
   for increasing periods of time between 1 and 20 min.
   Maximum fluorescence intensity has been reported with
   hydrolysis in 6N HCl at 20°C after 7 min. and in 1 HCl
   at 60°C after 11 min.

4. Rinse in water (5 min.).

5. Stain in freshly prepared BAO (2,5-bis 3-4 (4'-amino-
   phenyl- (1')) - 1,3,4-oxadiazol). The solution is made up
   as follows: 0.01% BAO in distilled water with 1 ml 1N HCl
   and 0.5 ml of a 10% NaHSO$_3$ solution. Shake and filter
   (2 hrs).

6. Wash in sulphite water (10 ml, 10% NaHSO$_3$ solution and
   10 ml 1N HCl in 200 ml distilled water). (3 times,
   2 min.).

7. Wash in water (10 min.).

8. Dehydrate in graded glycerol/water series.

9. Mount in glycerol.

10. Examine the fungal material as temporary squash prepara-
    tions in phase contrast transmitted light and in UV light
    (280-400 nm). Fluorescence emmission can be measured
    using a Zeiss photomicroscope and photomultiplier).

11. This technique is useful in the examination of DNA con-
    tents in nuclei of *Phytophthora*.

ORTIMER, A. MARTIN. 1975. Phytophthora Newsletter 22-24.

II. Acetocarmine technique for staining somatic and gametangial
    nuclei in *Phytophthora*   .

1. Incubate cultures of *Phytophthora* in 10-15 ml of cleared
   V-8 juice broth, in the dark at 25°C for 72 hrs.

2. Remove mycelial mats from the culture medium and rinse in
   deionized water.

3. Fix mycelial mats for 60 min. at 20°C in acid alcohol
   (1:3 glacial acetic acid: absolute ethanol).

4. Transfer to 40% acetic acid (HAC) for 5 min.

5. Stain with 1% acetocarmine solution (1 g Orth's Carmine
   Lithium in 40% HAC), thoroughly mixed with 4% iron alum.

6. After 15 min., remove stained mycelial fragments and
   rinse in 40% HAC to remove excess stain.

7. Mount in 40% HAC and heat gently.

8. Observe microscopically.

Note:

Prepare Acetocarmine stains by dissolving 1 g of Orth's Carmine
Lithium (Matheson, Coleman & Bell), in 100 ml 40% HAC in a
400 ml beaker. Reflux for 60-90 min. at moderate heat to dissolve the stain.

STEPHENSON, L.W., D.C. ERWIN and J.V. LEARY. 1973. Phytophthora Newsletter
1974. 2: 23-24.

IV. Technique for Phase Microscopy of Oospore Nuclei in *Phytop*
    thora

1. Incubate *Phytophthora* cultures in 10-15 ml cleared V-8
   juice broth in the dark at 25°C for 10 days.

2. Remove mycelial mats and rinse in deionized water.

3. Fix mycelial mats for 60 min. at 20°C in Herr's EPA_{50}
   fixative (5:5:90, 40% Formalin, Propionic Acid, 50%
   Ethanol).

4. Transfer fragments of mycelium to Herr's clearing agent
   (2:2:2:2:1, 85% lactic acid, chloral hydrate, phenol,
   clove oil and xylene w/w), in wells of No. 1 Coor's spo
   dish for 15 min.

5. Mount in Herr's clearing agent and observed with Zeiss
   Model RA microscope equipped with a 63x (NA 1.40) plana
   pochromatic phase contrast oil immersion objective.

Notes:

a) It is imperative that objectives of at least 1.2 NA be used.

b) Use caution in handling the clearing solution since it readily blisters skin and removes surface coatings from microscope stages.

STEPHENSON, L.W., D.C. ERWIN, and J.V. LEARY. 1973. Phytophthora Newsletter, 1974. 2: 22-23.

. Dehydration Sequence: Mycelium and gametangia

1. Place in solution (1) (2 changes)	1 hour
2. Transfer to solution (2)	1 hour
3. Transfer to solution (3)	2 hours
4. Transfer to solution (4)	12-18 hours
5. Transfer to solution (5)	1 hour
6. Transfer to solution (6)	1 hour
7. Transfer to solution (7)	12-18 hours

Solutions:

1) 30% ETOH☆

2) 50% ETOH

3) 40% (95%) ETOH, 10% N Butyl Alcohol, 50% H_2O prepared by mixing 40 ml 95% ETOH, 10 ml N Butyl and 50 ml of H_2O per each 100 ml of solution

4) 50% (95%) ETOH, 20% N Butyl Alcohol, 30% H_2O

5) 50% (95%) ETOH, 35% N Butyl Alcohol, 15% H_2O

6) 45% (95%) ETOH, 55% N Butyl Alcohol

7) Absolute ETOH

ETOH = Ethyl alcohol

For oospores increase time in (1-3, 5 & 6) to 24 hours and (4 & 7) to 48 hours each. If material is to be embedded additional transfer from (7) to N Butyl Alcohol for 12 hours and N Butyl + Eosin Y for 12 hours.

STEPHENSON, L.W. 1974. Dept. of Plant Pathology, University of California, Riverside, CA.

VI. Incorporation of charcoal for cytological studies

1. Make up the following medium: V-8 juice 50 ml/liter, pea extract 75 g/liter, 5 g/liter activated charcoal, 2% agar.

2. Inoculate with *Phytophthora*.

3. Oogonia and oospores produced on this medium appear almost colorless with the oogonial wall transparent. Struc-tures within the oogonial wall can be clearly discerned

LAVIOLA, C. 1974. Phytopath. Mediterran. 13: 55-58.

VII. Chromosome Numbers

P. *cactorum* . . . . n = 9 or 1

P. *cambivora* . . . . n =

P. *capsici* . . . . n = 6; n = 10 or 1

P. *cinnamomi* . . . . n = 9 or 1

P. *drechsleri* . . . . n = 9-12; n =

P. *erythroseptica* . . . . n = 9 or 1

P. *infestans* . . . . n = 9 or 1

P. *megasperma* var. *megasperma* . . . . n = 22-3

P. *megasperma* var. *sojae* . . . . n = 10-1

P. *nicotianae* var. *parasitica* . . . . n =

P. *palmivora* 's' type . . . . n =

P. *palmivora* 'L' type . . . . n =

Notes:

1) *P. megasperma* and *P. megasperma* var. *sojae* are believed to be tetraploid and diploid respectively.

2) Isolates of *P. infestans* from the British Isles are either polyploid or tetraploid.

MAIA *et al.* 1976. Annales Phytopathol. 8: 141-146.

SANSOME, E. 1977. J. Gen. Microbiol. 99: 311-316.

SANSOME, E. and C.M. BRASIER. 1974. Trans. Brit. Mycol. Soc. 63: 461-467.

STEPHENSON *et al.* 1974. Can. J. Botany 52: 2055-2060.

). Procedures for Fixing and Embedding Spores and Hyphae of *Phytophthora* for Thin Sectioning and Electron Microscopy

Hyphae, sporangia, zoospores, gametangia, and certain chlamydospores of *Phytophthora* can be chemically preserved and embedded utilizing standard techniques. The procedure listed below is not unique, but has been utilized, with slight modifications, in most studies dealing with the spores and hyphae of *Phytophthora* and other oomycetous fungi. The fixation and embedding of aged oospores and chlamydospores remains difficult. At this time no procedures have been established for consistently and adequately preserving these propagules for detailed ultrastructural study.

. Supplies and Chemicals

1. Glutaraldehyde, osmium tetroxide, sodium cacodylate, acetone, propylene oxide.

2. Spurr's resin, Epon 812, or other resin.

3. Small vials with plastic covers, pasteur pipettes, rubber bulbs, razor blades, EM forceps, embedding capsules and trays, plastic disposable beakers (100 and 400 ml), stir bars, wooden sticks, and 60°C oven.

Some U.S. companies which distribute E.M. supplies include:

Electron Microscopy Sciences   Ernest F. Fullam, Inc.
P. O. Box 251   P. O. Box 444
Fort Washington, PA 19034   Schenectady, NY 12301

Ladd Research Industries, Inc.   Ted Pella Company
P. O. Box 901   P. O. Box 510
Burlington, Vermont 05401   Tustin, California 92680

Tousimis Research Corporation
P. O. Box 2189
Rockville, Md. 20852

II. Prefixation Handling

Physical disruption of the specimens prior to fixation should
be avoided as much as possible. Violent spore isolation techniques and shearing forces may alter organelle distribution or
rupture hyphae or spores. Good preparations are obtained when
the culture is flooded with fixative after the growth medium
has been gently pipetted or decanted away. This technique requires that spores be sectioned individually, but avoids speci
men damage from adverse handling.

For fixing suspension of zoospores, fixative may be added by
drops to the swimming zoospores prior to centrifugation to con
centrate the spores.

Procedures for preparing hyphal tips for sectioning have been
described by H.C. Hoch, 1974, Preparation of fungal hyphae
grown on agar-coated microscope slides for electron microscopy
Stain Technol. 49: 318-320.

III. Fixation and Embedding

(Cautions: All glassware should be thoroughly clean. Fixatives
should be handled in the hood to avoid harmful vapors, Contact

of fixatives and resins with the skin should be avoided.)

1. 2% glutaraldehyde in 0.1 M sodium cacodylate buffer, pH
   7.2, at room temperature for one hour. (The pH of the
   buffer can be lowered to 7.2 by adding a few drops of
   1N HCl to the solution while stirring on a pH meter.)
   (Suspensions of zoospores, or other spores, can be fixed
   by adding the glutaraldehyde to the suspensions for one-
   half hour and then concentrated by centrifuging in coni-
   cal tubes at low speed for 5-10 minutes until a pellet
   is formed. The fixative is then decanted, the tube wiped
   dry to avoid dilution of the pellet, and a drop of molten
   agar equal to the volume of the pellet added. The agar
   and pellet must be stirred immediately with a bacterio-
   logical transferring loop and then touched with pasteur
   pipettes to draw up the agar-pellet suspension by capil-
   lary action. In a few seconds, as the agar hardens, the
   agar mixture can be blown out of the pipettes, cut into
   2-3 mm cylinders, and resuspended in glutaraldehyde.

2. Rinse in 0.1 M sodium cacodylate buffer, pH 7.2, with
   8-10 changes within 30 minutes.

3. Postfix in 1% osmium tetroxide in 0.1 M sodium cacodylate, pH 7.2, at room temperature for 1 hour.

4. Rinse in 0.1 M sodium cacodylate buffer, pH 7.2, with
   8-10 changes within 30 minutes.

5. Dehydrate in acetone: 30%, 50%, 70%, 90%, 95%, 100%
   (3 changes) - 10 minutes each. (During dehydration is
   a good time to prepare the resin. Resin kits contain
   instructions for proper mixing. Resins are hydroscopic
   and should be covered while mixing.)

6. Propylene oxide (3 changes) - 10 min. each.

7. 1:1 propylene oxide: resin - 1 hour.

8. 100% resin - overnight (approx. 16 hours).

9. Place specimens in fresh resin for one hour.

10. Place specimens in capsules with one drop of resin and
    place in oven at prescribed temperature for 1-2 hours.

(This prevents specimens from floating to the top of the capsule. Bubbles should be teased away from specimen.)

11. Fill capsules and leave in oven for prescribed time period.

IV. Selected References

1. HEMMES, D.E. and S. BARTNICKI-GARCIA. 1975. Ultrastructure of gametangial interaction and oospore development in *Phytophthora capsici*. Arch. Microbiol. 103: 91-112.

2. HEMMES, D.E. and L.D.S. WONG. 1975. Ultrastructure of chlamydospores of *Phytophthora cinnamomi* during development and germination. Can. J. Bot. 53: 2945-2957.

3. REICHLE, R.E. 1969. Fine structure of *Phytophthora parasitica* zoospores. Mycologia 61: 30-51.

4. BIMBONG, C.E. and C.J. HICKMAN. 1975. Ultrastructural and cytochemical studies of zoospores, cysts, and germinating cysts of *Phytophthora palmivora*. Can. J. Bot. 53 1310.

5. HOCH, H.C. and J.E. MITCHELL. 1972. The ultrastructure of *Aphanomyces euteiches* during asexual spore formation Phytopathology 62: 149-160.

6. GROVE, S.N. and C.E. BRACKER. 1970. Protoplasmic organization of hyphal tips among fungi: Vesicles and Spitzenkorper. J. Bact. 104: 989-1009.

7. LUNNEY, C.Z. and C.E. BLAND. 1976. Ultrastructure of mature and encysting zoospores of *Pythium proliferum* de Bary. Protoplasma 90: 119-137.

E. Some Ultrastructural Features of *Phytophthora*

1. Have continuous microtubules.

2. Have chromosomal microtubules.

3. The centriole arrangement is 180° (end to end alignment

4. Centriole migration occurs with concomitant spindle formation.

5. The nucleolus is persistent.

6. The dictyosome is believed to perform a dual role in *Phytophthora*: during direct germination, it appears to serve primarily as a transport system for precursor wall material: during indirect germination, it appears to function as a membrane donor.

7. Flagella form in both direct and indirectly germinating sporangia. The sporangium is thus not programmed to form either zoospores or germ tubes by virtue of a selective trigger mechanism, but rather always develops in the direction of zoospore formation, and then reverses the trend when conditions favor germ-tube formation.

HEMMES, D. and H. HOHL. 1973. Can. J. Botany 51: 1673.

HEMMES, D. and H. HOHL. 1969. Amer. J. Botany 56: 300-313.

E. Selected bibliography of ultrastructural aspects of zoospores, cysts, flagella, sporangia, chlamydospores, oospores, and haustoria of *Phytophthora*

I. Ultrastructure of Zoospores

1. HOHL, H. and S.T. HAMAMOTO. 1967. Ultrastructural changes during zoospore formation in *Phytophthora parasitica*. Amer. J. Botany 54: 1131-1139.

2. HO, H.H., K. ZACHARIAH, and C.J. HICKMAN. 1968. The ultrastructure of zoospore of *Phytophthora megasperma* var. *sojae*. Can. J. Botany 46: 37-41.

3. REICHLE, R.E. 1969. Fine structure of *Phytophthora parasitica* zoospores. Mycologia 61: 30-51.

4. DESJARDINS, P.R., M.C. WANG, and S. BARTNICKI-GARCIA. 1973. Electron microscopy of zoospores and cysts of *Phytophthora palmivora*: Morphology and surface textures. Archiv. Mikrobiol. 88: 61-70.

5. BIMBONG, C.E. and C.J. HICKMAN. 1975. Ultrastructural and cytochemical studies of zoospores, cysts, and germinating cysts of *Phytophthora palmivora*. Can. J. Botany 53: 1310.

6. SING, V.O. and S. BARTNICKI-GARCIA. 1975. Adhesion of *Phytophthora palmivora* zoospores: electron microscopy of cell attachment and cyst wall fibril formation. J. Cell Science 18: 123-132.

II. Ultrastructure of Cysts

1. HEMMES, D.E. and H.R. HOHL. 1971. Ultrastructural aspect of encystation and cyst-germination in *Phytophthora para sitica*. J. Cell Science 9: 175-191.

2. NOGUEIRA, M.L., P. PINTO DA SILVA, and S. BARTNICKI-GARCIA. 1977. Freeze-fracture and freeze-etching study of encystment of *Phytophthora palmivora* zoospores. J. Gen. Microbiol. 102: 149-155.

III. Ultrastructure of Flagella

1. FERRIS, V.R. and H.H. LYON. 1954. The flagella of swarm-spores of *Phytophthora infestans* as viewed with the phas and electron microscope. Phytopathology 44: 487.

2. DESJARDINS, P. R. *et al*. 1970. On the binucleate condition of the quadriflagellated zoospores of *Phytophthora palmivora*. Mycologia 62: 421-427.

3. HEMMES, D.E. and H.R. HOHL. 1972. Flagellum degeneration in the fungus *Phytophthora palmivora* (formerly *P. parasitica*). J. Gen. Microbiol. 73: 345-351.

4. DESJARDINS *et al*. 1973. Flagellar hairs on zoospores of *Phytophthora* species: Tip hairs on the whiplash flagellum. Experientia 29: 240-241.

IV. Ultrastructure of Sporangia

1. CHAPMAN, J.A. and R. VUJICIC. 1965. The fine structure
   of sporangia of *Phytophthora erythroseptica*. Pethyb. J.
   Gen. Microbiol. 41: 275-282.

2. KING, J.E., J. COLHOUN, and R.D. BUTLER. 1968. Changes
   in the ultrastructure of sporangia of *Phytophthora infestans* associated with indirect germination and ageing.
   Trans. Brit. Mycol. Soc. 51: 269-281.

3. HEMMES, D.E. and H.R. HOHL. 1969. Ultrastructural changes
   in directly germinating sporangia of *Phytophthora parasitica*. Amer. J. Botany 56: 300-313.

4. ELSNER, P.R. *et al*. 1970. Fine structure of *Phytophtora
   infestans* during sporangial differentiation and germination. Phytopathology 60: 1765-1772.

5. WILLIAMS, W.T. and R.K. WEBSTER. 1970. Electron microscopy of the sporangia of *Phytophthora capsici*. Can. J.
   Botany 48: 221-227.

6. CHRISTEN, J. and H.R. HOHL. 1972. Growth and ultrastructural differentiation of sporangia in *Phytophthora palmivora*. Can. J. Microbiol. 18: 1959-1964.

7. HEMMES, D.E. and H.R. HOHL. 1973. Mitosis and nuclear
   degeneration: Simultaneous events during secondary
   sporangia formation in *Phytophthora palmivora*. Can. J.
   Botany 51: 1673-1675.

8. HEMMES, D.E. and H.R. HOHL. 1975. Ultrastructural changes
   and a second mode of flagellar degeneration during ageing
   of *Phytophthora palmivora* sporangia. J. Cell Science 19:
   563-577.

. Ultrastructure of Gametangia

1. VUJICIC, R. 1971. An ultrastructural study of sexual reproduction in *Phytophthora palmivora*. Trans. Brit. Mycol.
   Soc. 57: 525-530.

2. HEMMES, D.E. and S. BARTNICKI-GARCIA. 1975. Electron

microscopy of gametangial interaction and oospore development in *Phytophthora capsici*. Archiv. Mikrobiol. 103: 91-112.

3. BARTNICKI-GARCIA, S. and D.E. HEMMES. 1975. Some aspects of the form and function of Oomycete spores. *In*, The Fungal Spore: form and function (Hess, W.M. & D.J. Weber eds.). John Wiley & Sons. N.Y.

4. HEMMES, D.E. and O.K. RIBEIRO. 1977. Electron microscopy of early gametangial interaction in *Phytophthora megasperma* var. *sojae*. Can. J. Botany 55: 436-447.

VI. Ultrastructure of Chlamydospores

HEMMES, D.E. and L.D.S. WONG. 1975. Ultrastructure of chlamydospores of *Phytophthora cinnamomi* during development and germination. Can. J. Botany 53: 2945-2957.

VII. Ultrastructure of Hyphae and Haustoria

EHRLICH, M.A. and H.G. EHRLICH. 1966. Ultrastructure of the hyphae and haustoria of *Phytophthora infestans* and hyphae of *P. parasitica*. Can. J. Botany 44: 1495-1503.

Section 13:

Cell wall composition

C. Cell Wall composition

Oomycetes (*Phytophthora*) cell walls have a unique chemical composition that differs from other Phycomycetes in having cellulose rather than chitin as the main structural component.

In *Phytophthora*, glucose makes up 80-90% of the dry weight of
the cell wall. The β-glucans, comprising the glucose component
are of two kinds: (1) cellulose, consisting of characteristic
-1,4 linked polymers and, (2) insoluble branched glucans, consisting mainly of β-1,3 links and β-1,6 linkages.

Table 10 gives the chemical composition of glucose, protein,
lipid, amino acids, and phosphorus of various structures in the
life cycle of *Phytophthora*.

Table 10: Chemical Composition of the Walls of *Phytopthora*

Component	Hypha 1[a]	Hypha 2[b]	Sporangium 3[c]	Cyst 3[c]	Oospore/Oogonium 4[d]
Hexose[f]	88.0	86.0	93.5	92.5	77.7
Protein	3.5	4.2	1.7	2.4	10.66
Lipid: Free	0.3	0.2	e	e	5.68
Bound	2.1	0.9	e	e	5.64
Amino sugar	0.3	0.3	0.2	0.1	0.43
Phosphorus	0.13	e	0.05	0.098	0.55

Phytophthora cinnamomi, from Bartnicki-Garcia (1966)
Phytophthora parasitica, Ibid.
Phytophthora palmivora, from Tokunaga and Bartnicki-Garcia (1971b).
Phytophthora megasperma var. *sojae*. (Lippman *et al.*, 1974).
Not measured.
Acid hydrolysis showed that glucose was the main hexose (> 95%); mannose was
also present.

The amino acids reported to be present in *Phytophthora* cell
walls are given in Table 11.

Table 11: Overall amino acid composition of protein from hyphal walls of
   Phytophthora cinnamomi

	% of total amino acids
Threonine	14.8
Glutamic Acid	11.1
Serine	9.8
Aspartic Acid	9.0
Alanine	9.0
Proline	5.5
Hydroxyproline	5.0
Valine	5.0
Lysine	4.7
Leucine	4.5
Glycine	4.5
Arginine	4.2
Ammonia	3.7
Tyrosine	2.6
Phenylalanine	2.1
Isoleucine	1.8
Histidine	1.1
Cysteine	1.1
Methionine	1.1

Note: The hyphal walls of *P. cinnamomi* and *P. parasitica* differ only
   slightly in quantitative composition.

Reproduced from S. BARTNICKI-GARCIA. Chemistry of hyphal walls of *Phytophthora*. J. Gen. Microbiol 42: 57-69. By permission of Cambridge Univ. Press.
Copyright 1966.

The early cytochemical observations of Mangin (1895), Dastur
(1913), and Blackwell (1943) indicated that cellulose was pres
ent in the oospore and oogonial walls of *Phytophthora*.   These
observations were confirmed by the elegant biochemical studies
of Lippman *et al.* (1974). See Table 10.

The amino-acid composition of the oospore-oogonial wall is
shown in Table 12.

Table 12: Amino acid composition of the oospore-oogonial wall proteins of
P. megasperma var. sojae

	Mole (%)	Wall dry wt. (%)
Alanine	10.14	0.49
Arginine	9.95	0.94
Aspartic	9.35	0.67
Glutamic	8.15	0.65
Glycine	20.88	0.85
Histidine	0.48	0.04
Hydroxyproline	3.66	0.26
Isoleucine	1.41	0.10
Leucine	4.22	0.30
Lysine	3.15	0.25
Methionine	0.74	0.06
Phenylalanine	1.79	0.16
Proline	5.61	0.35
Serine	7.37	0.42
Threonine	7.73	0.50
Tyrosine	2.54	0.25
Valine	2.83	0.18
	100.00	6.48

Reproduced from LIPPMAN et al. Isolation and chemical composition of oospore-
oogonium walls of Phytophthora megasperma var. sojae. J. Gen. Microbiol. 80:
31-141. By permission of Cambridge Univ. Press, Copyright 1973.

Notes:

1. There is an increase in lipid and protein in oogonial-
   oospore walls, compared to that found in mycelial walls.

2. Glycine and arginine are the most abundant components of
   the amino-acids in oogonial-oospore walls; in mycelial
   walls, threonine, glutamic acid, and serine are the prin-
   cipal amino-acids.

3. Phosphorus content (0.55%) of oogonial-oospore walls is
   much higher than that found in hyphal walls and in sporangia.

. There are differences in the proportions of C-3 and C-4
   linked residues of the glucans in various stages in the life
   history of Phytophthora (e.g. cysts, hyphae etc.). The
   branching of the C-3 and C-6 linked residues also differ.
   (Tokunaga and Bartnicki-Garcia, 1971, Arch. Mikrobiol. 79:
   293-310.)

II. Storage-polysaccharides

The storage-polysaccharide in *Phytophthora* is a soluble β-1,3 -
β-1,6 linked D-glucan known as mycolaminaran. Glycogen in not
present in *Phytophthora*. Mycolaminaran is made up of glucose
units with the linear chains linked by β 1-3 bonds and one or
2 branches at C-6, joined by 1,6 β-D-linkages. The developmenta
stage of the fungus influences the number of branches and the
molecular size of mycolaminaran. (Wang, M.C. and S. Bartnicki-
Garcia, 1974. Carbohydrate Res. 37: 331-338.) The quantitative
changes in mycolaminaran during the development of various
stages in the life cycle of *Phytophthora* are illustrated in
Figure 7.

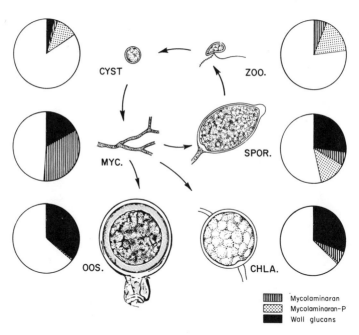

Fig. 7: Changes in β-glucan content during the life cycle of *Phytophthora*.
Circles represent total dry weight. Zoospores, cycts, mycelium and sporang
of *P. palmivora*; oospores of *P. megasperma* var. *sojae*; chlamydospores of *P*
parasitica. By permission from BARTNICKI-GARCIA and HEMMES *in* The Fungal
Spore: Form and Function (D.J. Weber & W.M. Hess, eds.). John Wiley & Sons
Inc. Copyright 1976.

References:

BARTNICKI-GARCIA, S. 1966. J. Gen. Microbiol. 42: 57-69.

BARTNICKI-GARCIA, S. 1968. Annu. Rev. Microbiol. 22: 87-108.

BARTNICKI-GARCIA, S. and D.E. HEMMES. 1976. Some aspects of the form and
function of Oomycete spores, pages 594-640. *In*, The Fungal Spore: Form and
function (D.J. Weber & W.M. Hess, eds.). John Wiley & Sons, Inc. New York.

LIPPMAN, E., D.C. ERWIN, and S. BARTNICKI-GARCIA. 1974. J. Gen. Microbiol.
80: 131-141.

Useful general references:

BARTNICKI-GARCIA, S. 1970. Cell wall composition and other biochemical
markers in fungal phylogeny, pages 81-103. *In*, Phytochemical Phylogeny (J.
B. Harborne, ed.). Academic Press. New York.

BARTNICKI-GARCIA, S. 1973. Fungal cell wall composition. *In*, Handbook of
Microbiology, Vol. 2 (A. I. Laskin and H.A. Lechevalier, eds.). Chemical
Rubber Press. Cleveland, U.S.A.

Section 14:

Miscellaneous techniques

A. Purification of contaminated *Phytophthora* cultures

C. 3P Medium

1. Prepare a lima bean medium (see page 76).
2. After autoclaving add:

   Penicillin G . . . . 100 ppm

   Polymyxin . . . . 100 ppm

   Pimaricin . . . . 100 ppm
3. Pour into petri plates.
4. Grow contaminated culture on above medium. Bacterial colonies are sufficiently inhibited to allow clean hyphal tips to be transferred to a normal culture medium.

GALLEGLY, M.E. 1960. West Virginia Univ. Dept. of Plant Pathology.

I. "Pancake" Method

1. Prepare the following medium: per liter

   Dextrose . . . . 10 g

   $NH_4H_2PO_4$ . . . . 2 g

   KNO_3 . . . . 1 g

   $MgSO_4$ . . . . 1 g

   Distilled water . . . . 1000 ml

   Agar . . . . 20 g
2. Autoclave the above medium and pour in petri plates to give a layer of medium at least 5 mm in thickness.
3. Divide the solidified medium in each plate into 4 sectors using a sterile spatula.
4. Inoculate each sector with a piece of mycelium taken from a contaminated culture.
5. Lift 3 of the sectors out of the petri plate with a sterile spatula and place *inverted* in separate petri plates

so that the inoculum is completely covered by the agar block. The 4th sector is inverted in the original dish.

6. Incubate at 20-25°C.

7. Mycelium of *Phytophthora* will grow up through the agar and spread on the surface. Bacterial contamination is confined to the lower layer of agar.

8. This method can also be used to isolate *Phytophthora* from diseased tissue or soil.

SLEETH, B. 1945. Phytopahtology 35: 1030-1031.

Note:

For severely contaminated cultures, we have found that this method in combination with an antibiotic method (see pages 57-63), can be very effective.

III. Van Tiegham cells

1. Mold two or three small round pellets of clay (2-3 mm i diameter), and place on one end of a micro slide ring.

2. Place the ring in the bottom part of a petri dish with the pellets touching the glass.

3. Press the ring gently until a clearance of approximatel 2 mm from the bottom is obtained.

4. Replace the cover of the petri dish and autoclave the whole apparatus.

5. After sterilization, cool and pour a suitable melted ag medium into the petri dish so as to fill the dish   to within 3-5 mm from the upper edge of the inside of   the ring.

6. After the agar has solidified, place a small piece of t contaminated material on the agar in the middle of   the ring.

7. The fungus hyphae tend to grow into the agar and come c

from under the ring to the surface of the agar outside
the ring, free from any bacteria.

8. The agar can also be acidified with a few drops of 25%
lactic acid to retard bacterial growth.

Advantages:

Several rings can be mounted in one plate. The agar holds the
rings firmly in place so plates can be inverted safely.

CLARK, P.A. and R.S. DICKEY. 1950. Phytopathology 40: 389-390.

V. Inverted agar dilution plate method

1. Obtain zoospores from contaminated *Phytophthora* culture.

2. Place drops of spore suspension on the surface of V-8
juice agar and disperse by adding a few drops of sterile
distilled water.

3. Invert the entire agar disc with a spatula.

4. Remove the convex rim of agar at the side of the plate
so that the agar surface adheres to the bottom of the
plate without entrapment of air.

5. Spores can be examined through the bottom of the plate.

6. The growth of bacteria and actinomycetes are generally
confined to the agar-glass interface, while the mycelium
from germinated spores grows up through the agar.

7. Pure culture transfers from the top surface of the agar
can be made by taking shallow agar plugs.

8. The inverted-agar dilution-plate technique is a rapid
method for isolation of single-spore cultures from a
variety of fungi.

Note:

Bacteria in suspect cultures can be detected by transferring agar and mycelium to Eugon broth. If bacteria are present, the broth will become turbid in 24-48 hrs.

SCHMITTHENNER, A.F. and J.W. HILTY. 1962. Phytopathology 52: 582-583.

B. Long Term Storage of *Phytophthora* Species

Since *Phytophthora* spp. do not readily form resting cells, maintenance of isolates in culture is difficult unless frequent ly sub-cultured. However, for long term storage, the following methods have been used successfully by various investigators.

    Method I: Mineral Oil

    1. Use young *Phytophthora* cultures (3-5 days old), growing
       on lima bean or V-8 juice agar slants. (20-25 ml. test-
       tubes.)

    2. Completely submerge culures with sterile mineral oil.

    3. Store at 10-15°C for 1 1/2 - 2 yr. before transfer.

    4. *P. infestans* should be transferred every 9-12 months.

Note:

This method has proved to be adequate for the world collection of *Phytophthora* spp. maintained by the departments of plant pathology at West Virginia University, Morgantown, W. Va. (M.E Gallegly), and at the University of California, Riverside, Ca. (G.A. Zentmyer).

See also:

WERNHAM, C.C. 1946. Mineral oil as a fungus culture preservative. Mycologi 38: 691-692.

WERNHAM, C.C. and H.J. MILLER. 1948. Longevity of fungus cultures under mineral oil. Phytopathology 38: 932-934.

Method II: Hemp Seed in Distilled Water

1. Place 3 or 4 hemp seeds in 100 ml distilled water in
   200 ml screw cap prescription bottles.

2. Autoclave at 15 psi for 20 min.

3. Place small pieces of agar on which the fungus is growing, into the bottle and tighten cap.

4. Recovery of *Phytophthora* is possible after 1-4 yr.

5. Addition of 10 ppm benomyl [methyl 1-(butyl-carbamoyl)-
   2-benzimidazole carbamate] to the transfer medium aids
   in the recovery of *Phytophthora* from contaminated cultures.

AABE, R.D., J.H. HURLIMAN and R. EMERSON. 1973. 2nd. Intl. Pl. Path. Con-
ress. Minn. (Abstr.).

Method III: Low Temperature Storage

1. Grow fungi at $24^{O}C$ in petri plates containing 30 ml of
   the following medium:

   $CaCl_2$ . . . . 0.05 g

   NaCl . . . . 0.025 g

   KH_2PO_4 . . . . 0.5 g

   $(NH_4)_2HPO_4$ . . . . 0.25 g

   $MgSO_4 \cdot 7H_2O$ . . . . 0.15 g

   $FeCl_3$ (1%) . . . . 1.2 ml

   Thiamine HCl . . . . 100 μg

   Malt extract (paste) . . . . 3 g

   Glucose . . . . 10 g

   Agar . . . . 15 g

   Distilled water . . . . 1000 ml

   pH should be between 5.5-5.7 after autoclaving.

2. Sterilize 2.5 x 15 cm capped Pyrex tubes containing 25 ml
   sterile distilled water.

3. When cultures are approx. 40 cm in diameter, cut 8 mm
   diameter mycelial discs from the colony with a sterilized cork borer.

4. Aseptically remove 10-15 discs and place in the sterilized tubes.

5. Store all tubes in darkness at 5°C.

6. Isolates stored in this manner did not lose virulence.

7. Survival rate of *Phytophthora* drops rapidly after 1 year
   of storage.

8. This method is claimed to be better than storage on agar
   slants which require transfer every 3-4 months with considerable loss in virulence.

MARX, D.H. and W.J. DANIEL. 1976. Can. J. Microbiol. 22: 338-341.

Method IV: Distilled Water

1. Half-fill screw capped bottles (90 x 26 mm) with distilled water, autoclave with loosened cap and cool to
   room temperature.

2. Cut blocks approx. 60 mm^3 from the edge of fungal cultures growing on agar.

3. Transfer blocks into the sterile bottles and screw tightly to seal the bottles.

4. Store at room temperature.

5. *Phytophthora* spp. have been stored without change in
   viability or morphology when tested seven years later.

BOESEWINKEL, H.J. 1976. Trans. Brit. Mycol. Soc. 66: 183-185.

Method V: Liquid Nitrogen

1. Place short pieces of lima bean epicotyl infected with
   downy mildew in a sterile 2 or 5 ml polypropylene serum
   test-tube fitted with a screw cap and silicone washer.

2. Place the sealed tube directly in a vapor phase storage
   liquid-nitrogen refrigerator (-100°C to -160°C).

3. When removed from the refrigerator, allow the tube to
   thaw at room temperature (10 min.), and immediately use
   inoculum in lima bean resistance screening tests.

4. Tests indicate that field inoculum of *P. phaseoli* is
   virulent after storage in liquid N for 12 months.

dvantages:

) No protective agent is required in the tubes.

) The rate of freezing is not critical.

AN ANTONIO, J.P. and V. BLOUNT. 1973. Plant Dis. Reptr. 57: 724.

. Control of Mite Infestations

. Gelatin-CuSO$_4$ Technique.

he presence of mites in fungus cultures can be a serious prob-
em, particularly if infestations occur in stock cultures. Once
nfested, *Phytophthora* cultures are very difficult to clean. It
s best therefore, to routinely follow a good preventative pro-
ram. The following method (Snyder & Hansen) has proved to be
ffective for all our stock cultures.

1. Make up a 20% soln. of gelatin in water.

2. Add 2% CuSO$_4$ to the above solution to prevent fungal and
   bacterial growth.

3. Obtain a book of cigarette papers - preferably the L.L.F.
   brand (Riz la Croix) and a piece of heavy blotting paper.

4. Pour 25 ml of the melted gelatin/CuSO$_4$ mixture into a
   petri dish and allow to solidify.

5. Take the cigarette papers from their cover, trim off the
   glue that holds the papers together and cut the bundle
   of sheets in half.

6. Place cut sheets in a petri dish and sterilize in a dry
   oven.

7. This dry heat treatment makes the sheets separate more
   easily at time of sealing a tube.

8. After an agar slant test-tube has been inoculated in the
   usual manner, push the cotton plug down inside the tube.

9. Flame the rim of the tube.

10. Hold tube upside down and press the flamed rim gently
with a rotary motion against the surface of the solidified gelatin until the rim is coated with a thin film of
melted gelatin.

11. Place the gelatin-coated rim against the cigarette paper
in the petri dish so that the top sheet adheres to the
rim and is thus neatly picked up.

12. Press rim + paper firmly against the blotting paper to
ensure that the paper will adhere tightly to the rim of
the tube.

13. Place the tube upright in a rack with other tubes simi-
larly prepared so that the corners of the projecting
pieces of paper touch each other.

14. Ignite the papers at a single point; the projecting paper on all tubes will burn off, leaving circular paper
seals that effectively keep out all mites, spores and
other contaminants.

15. Burn off seals when sub-cultures are required and reseal
when transfers are completed.

SNYDER, W.C. and H.N. HANSEN. 1946. Mycologia 38: 455-462.

For several less effective but adequate techniques, see:

1. CROWELL, I.H. 1941. Use of dichloricide in the control
of scavenger mites in test-tube cultures. Mycologia 33:
137.

2. PEASE, D. 1937. The insect menace in the bacteriologica
laboratory. Jour. Bact. 33: 619-624.

3. SHAFIK, M. and A.B.P. PAGE. 1930. Control of mites attacking stocks of insect and fungus cultures. Nature 12
311-312.

II. Mite Control (Water Barrier)

1. Place wire baskets of cultures on a pedestal in a pan o
water.

Note:

Method is effective against pedestrian mites but not against
those carried on the hands or clothing of the worker or those
carried by flies and other winged insects.

2. If a single tube happens to be infested, all cultures
   within the barrier will soon be invaded.

BARNES, B. 1933. Trans. Brit. Mycol. Soc. 18: 172-173.

II. Mite Control (Pyridine)

1. Invert a large bell-jar (approx. 20 liters capacity) and
   place a flat dish containing about 20 ml of commerical
   pyridine, covered by a wire gauze.

2. Place infested cultures, without removing the cotton
   plugs inside the bell-jar for 16 hrs (overnight).

3. Close the jar with a glass plate and seal with modeling
   clay or plasticine.

4. Subcultures after the above treatment should be free from
   mites.

5. In cases of severe infestation, or in very cold weather,
   give two exposures of 16 hrs each separated by a period
   of 14 to 16 days to allow unkilled eggs to hatch.

Note:

Due to the disagreeable odor of pyridine, it is best to carry
out treatments in a fume hood.

NEWSON, S.T. and F. TATTERSFIELD. 1922. Ann. Appl. Biol. 9: 213-240.

D. Media for pigment production of *Phytophthora* spp.

I. Timmer's Medium   per liter

1. $MgSO_4 \cdot 7H_2O$ . . . . 0.10 g

    K_2HPO_4 . . . . 1.04 g

    KH_2PO_4 . . . . 1.90 g

    $FeSO_4 \cdot 7H_2O$ . . . . 1 mg

    $CaCl_2 \cdot 2H_2O$ . . . . 1 mg

    Thiamine-HCl . . . . 1 mg

    Casein Hydrolysate . . . . 6 g

    Tyrosine . . . . 1 g

    Distilled water . . . . 1000 ml

    1 ml of a microelement solution containing Zn^{2+}, 1000ppm Cu^{2+}, 20 ppm; Mo^{6+}, 20 ppm; Mn^{2+}, 20 ppm.

    Agar . . . . 15 g

2. Sterilize at $121^{O}C$ for 15-20 minutes.

Notes:

a) Glucose inhibits pigment production.

b) Pigment production is not a useful taxonomic criterion sin
    pigmentation varies considerably in the progeny.

TIMMER *et al.* 1970. Mycologia 62: 967-977.

II. Shepherd's Medium   per lite

1. Oxoid casein hydrolysate (acid) 4.0

    L-tyrosine 1.0

    Potassium dihydrogen phosphate 2.0

    Magnesium sulfate 0.5

Calcium chloride . . . . 0.1 g

Thiamine hydrochloride . . . . 2.0 mg

Cholesterol . . . . 20 mg

Trace element solution . . . . 1.0 ml

Agar . . . . 18.0 g

Distilled water . . . . 1000 ml

The trace elements contain: Fe^{3+} (as the chloride or as sequestrene (sodium ferric diethylene triamine pentaacetate), 1 mg; Zn^{2+}, 1 mg; Cu^{2+}, 100 mg; Mn^{2+}, 10 μg; and Mo^{6+} as $(NH_4)_6Mo_7O_{24} \cdot 4H_2O$, 10 μg.

2. Adjust pH to 6.5 before autoclaving.

SHEPHERD, C.J. 1976. Aust. J. Botany 24: 607-617.

3. Mating Types Determination of 'maleness' and 'femaleness' of A1 and A2

1. Grow isolates for pairing on V-8 juice medium in petri plates.

2. Blocks containing mycelium are removed from the plates and placed on L-shaped strips of aluminum foil previously mounted on a microscope slide.

3. A block of sterile water agar is placed between the aluminum foil strips through which rows of minute holes of needle-point size have been pricked.

4. Push the three parts together and incubate 2-3 days in a moist chamber at 20°C.

5. The sparse growth on the water agar block permits tracing of the hyphae from gametangia back to their origin.

6. "Maleness" or "femaleness" of strains can thus be ascertained.

For best results:

1. The agar blocks should have a thickness no more than

4 mm and contain the fungus taken from vigorously growing colonies.

2. Not more than 10 holes should be made in the aluminum
   foil.

3. Arrange the holes in two rows; the upper row at about
   the same level as the upper surface of the agar block;
   the lower row at almost the level of the lower surface
   of the block. This forces sexual body formation close to
   the upper or lower surfaces of the agar block, making
   observations with the 10X objective feasible.

4. The petri dish moist chambers, as well as the microscope
   slides and aluminum strips are always autoclaved prior
   to use.

GALINDO, A.J. and M.E. GALLEGLY. 1960. Phytopathology 50: 123-128.

F. Staining *Phytophthora* oospores in roots

Bromphenol blue (100 mg) + AgNO$_3$ (3 g) in 50 ml of 95% ethanol
has been used to successfully stain oospores of *P. megasperma*
var. *sojae* in soybean roots. (Slusher, R.L. and J.B. Sinclair.
1973. Phytopathology 63: 1168-1171).

Oospores can readily be observed in soil and roots when stained
with a 1% solution of acridine orange, or 100-300 ppm Calcofluor, and viewed with fluorescence microscopy.

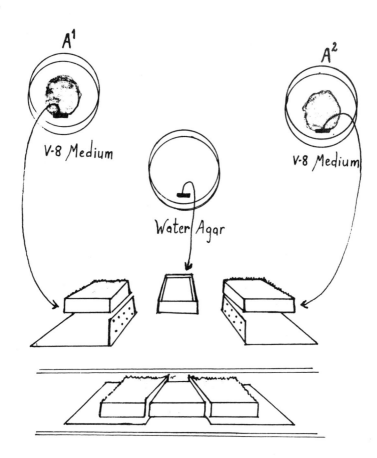

ig. 8: Diagram of the materials and method used in tracing gametangial
yphae of P. *infestans*. Blocks of V-8 medium containing mycelium of A1 and
2 types were removed from petri plates and placed on L-shaped strips of
luminum foil bearing lines of tiny holes above and below. A block of water
gar was placed between the aluminum strips, and the three parts, mounted
n a microscope slide, were pushed together. Redrawn by permission.

Section 15:

Other information pertaining to

Phytophthora

.. The uniqueness of *Phytophthora*

'he following characteristics are unique to the Pythiaceae only.

1. Cellulose and beta-glucans as the main cell wall poly-
   saccharide constituents, instead of chitin.

2. Biflagellate zoospores.

3. Oospores enveloped in a resistant, protective oogonium
   wall.

4. Sensitivity to certain antibiotics, e.g. chloramphenicol.

5. Resistant to polyene antibiotics (e.g. pimaricin, polymyxin, vancomycin, etc.).

6. Inability to synthesize sterols.

7. Non-requirement of sterols for vegetative growth, except
   for *P. megasperma.*

8. Requirement of exogenous sterols for sporulation (chlamydospores, oospores, etc.).

9. Diploid vegetative life-cycle.

10. Presence of microbodies.

11. Complex endoplasmic reticulum and dictyosome system as in
    higher organisms.

12. Method of fertilzation--female penetrates male.

13. All described species are pathogens of higher plants
    (this characteristic is not confined to the genus *Phy-
    tophthora*).

14. *Phytophthora* offers unique advantages for multi-disci-
    plinary research. The life-cycle includes such varied
    structures as mycelium, zoosporangia, swimming zoospores,
    encysted zoospores, chlamydospores, and oospores.

IBEIRO, O.K. Dept. of Plant Pathology, Univ. of Calif., Riverside.

SAO, P.H. Dept. of Plant Pathology, Univ. California, Riverside.

B.   Chemotaxis

Goode (1956) reported that zoospores of *P. fragariae* accumulate
behind the root tip of strawberry, and on germination, the resulting germ tubes grew towards the root tip, penetrating and
infecting the host plant. It has now been shown that a wide
range of chemicals can induce zoospore accumulation (chemotaxis
including amino acids, carbohydrates, ethanol, and inorganic
acids. Synergism between amino acids, ethanol, and sucrose has
been demonstrated by Halsall (1976). Amino acids appear to be
the strongest chemotactic attractants for zoospores. The most
chemosensitive amino acids are aspartic and glutamic acids,
with lesser chemotactic responses being recorded for arginine
and methionine. Structural requirements for attraction are reported to include the α-amino-acid group with a short carbon
chain terminating in an amide group (Halsall,   1978).

Zoospores show different degrees of attraction, depending on the
pH of the environment. At pH 3.0, zoospores show greater attraction to glutamate and aspartate, than at pH 5.0.

Ethanol, which can possibly arise from anaerobic fermentation by
roots under waterlogged conditions, has elegantly been demonstrated by Allen and Newhook (1975), to strongly attract zoospores. Their studies have led them to construct a 'rhizosphere
model' for establishing fixed concentration gradients of chemotactic agents.

Negative chemotaxis of *P. cinnamomi* in association with hydro-
chloric acid and its salts has been demonstrated by Allen   and
Harvey (1974).

Zoospores of *Phytophthora* species are now known to be attracted
to roots of both hosts and non-host plants. In some specific
cases however, zoospores have been found to be attracted to
roots of the host plant only, e.g. *P. cinnamomi* zoospores show
specificity to roots of avocado (*Persea americana*) only (Zentmyer, 1966).

Chemotaxis thus appears to be a significant phenomenon in path

enesis of *Phytophthora* to roots of host as well as non-host
lants.

LLEN, R.N. and F.J. NEWHOOK. 1973. Trans. Brit. Mycol. Soc. 61: 281-302.

LLEN, R.N. and J.D. HARVEY. 1974. J. Gen. Microbiol. 84: 28-38.

UKES, P.D. and J.L. APPLE. 1961. Phytopathology 51: 195-197.

OODE, P.M. 1956. Trans. Brit. Mycol. Soc. 39: 367-377.

ALSALL, D.M. 1976. Can. J. Microbiol. 22: 409-422.

ICKMAN, C.J. 1970. Phytopathology 60: 1128-1135.

ICKMAN, C.J. and H.H. HO. Annu. Rev. Phytopathol. U.S.A. 4: 195-200.

HEW, K.L. and G.A. ZENTMYER. 1973. Phytopathology 63: 1511-1517.

OUNG et al. 1977. Trans. Brit. Mycol. Soc. (In Press).

ENTMYER, G.A. 1966. Phytopathology 56: 907.

ote:

technique for demonstrating accumulation of zoospores of *Phy-
ophthora* on roots in soil has been described by R.S. Mehotra,
an. J. Botany 48: 879-882 (1970).

. Electrotaxis

oospores are attracted to the anode at currents of less than
.5 µA (less than 1.2 v/cm). Repulsion from the anode occurs at
igher currents. However, it has been reported that zoospores
an migrate to either the anode or cathode depending on the
ypes of organic acids present in the medium. Ho and Hickman
Can. J. Botany 45: 1963-1981, 1967), found that zoospores of
. *megasperma* var. *sojae* were not attracted to either of

the poles when an electric field was applied.

Studies also indicate that the zoospore surface is negatively
charged (Khew and Zentmyer, 1974; Halsall, 1976).

References:

KHEW, K.L. and G.A. ZENTMYER. 1974. Phytopathology 64: 500-507.

KATSURA et al. 1966. Ann. Phytopathol. Soc. Japan. 32: 215-220.

TROUTMAN, J.L. and W.H. WILLS. 1964. Phytopathology 54: 225-228.

D. Rheotaxis

Zoospores have been reported to respond to changes in localized
water currents (Katsura, K. and Y. Miyata. 1966. Kyoto Prefec-
tual Univ. Sci. Rept. 18: 51-56.

E. Negative Geotaxis

Evidence for the accumulation of P. cinnamomi motile zoospores
near the top surface of liquids is presented by (Palzer, C. 19
Phytophthora Newsletter 3: 9-11).

Useful Reference:

GOODAY, G.W. 1975. Chemotaxis and chemotropism in fungi and algae, pages
155-204. In, Primitive Sensory and Communication Systems, (M.J. Carlile,
ed.). Academic Press. New York.

F. Effects of light on Phytophthora

    1. Sporangium production
       Near-UV radiation (320-400 nm) from 8 to 100 μW cm^{-2}

significantly enhances sporulation in *P. capsici* and *P. palmivora* isolates.

P. cinnamomi and *P. megasperma* do not produce sporangia with exposure to light alone (see pages 111 & 120 for methods to induce sporangium production in these species). However, sporangium production is significantly enhanced in *P. megasperma* if cultures are exposed to continuous light after they have been placed in a sporulation medium. *P. cinnamomi* on the other hand, does not show a significant response to radiation from any wavelength between 320-1300 nm, when cultures are placed in a sporulation medium. The intensity of irradiation for these tests was between 100 and 200 μW cm^{-2}. At very low light intensities (8 μW cm^{-2}), *P. cinnamomi* does show an enhancement in sporangium production, but only in the near-UV.

2. Zoospore differentiation and release

Significantly fewer sporangia produced under near-UV radiation differentiate and release zoospores. The percentage of sporangia releasing zoospores varies with the isolate and the species (see Ribeiro *et al.*, Mycologia 68: 1162-1173, 1976).

3. Oospore production

Oospore production is significantly higher in the far-red wavelengths (750-850 nm) at 8 μW cm^{-2}. At 100 μW cm^{-2}, oospore production is less than that obtained in darkness.

4. Oospore germination

See section 6B, page 163.

complete bibliography on the effects of light on *Phytophthora* given by Ribeiro *et al.* (1976).

ferences:

BEIRO, O.K., G.A. ZENTMYER, and D.C. ERWIN. 1975. Phytopathology 65: 904-7.

RIBEIRO, O.K., G.A. ZENTMYER, and D.C. ERWIN. 1976. Mycologia 68: 1162-1173

ZENTMYER, G.A. and O.K. RIBEIRO. 1977. Phytopathology 67: 91-95.

Note:

Interactions between light and natural media, e.g. V-8 juice, lima bean, carrot agar, etc., are known to occur. These interactions are often detrimental to sporangia formation. The idea culture medium for studying the effects of light on *Phytophthora* is one in which the ingredients are all chemically defined, and do not contain any pigments that are photo-sensitive, e.g. Ribeiro's or Bartnicki-Garcia's synthetic media.

5. Light equipment

   a. For exposure of cultures at very low light intensities, see the apparatus described by Poff and Norris (1967), and Ribeiro *et al.* (1975). The modified monochromatic filters described in these two publications are available from Carolina Biological Supply, Burlington, North Carolina 27215, U.S.A. The four filters available are CBS Blue (450 nm), CBS Green (545 nm), CBS Red (650 nm), and CBS Far-Red (750 nm). These filters can be obtained as 4" or 12" squares.

   b. For exposure of cultures to higher intensities, colored fluorescent lamps can be used. The modification required to block off extraneous light from other wavelengths, is described by Ribeiro *et al.* (1976).

      Spectra of fluorescent lamps commonly used to irradiate cultures are given in Appendix II. Full spectra lamps are available from Verd-A-Ray Corp. 615 Front Street, Toledo, Ohio 43605, U.S.A. (Indor Sun lamps, guaranteed for 24,000 user hours), and Duro-Test Cor North Bergen, N.J. 07047, U.S.A. (Vita-Lite, guaranteed for 36,000 user hours).

   c. An apparatus for the automatic exposure of cultures to radiation has been described by Leach (Radiation Botany 2: 1-6, 1962).

d. Descriptions of chemical filters suitable for transmission of certain wavelengths in the UV and near-UV
are given by Wladimiroff (Photochem. Photobiol. 5:
243-250, 1966), and Muel and Malpiece (Photochem.
Photobiol. 10: 283-291, 1969).

6. Light measurements

a. UV measurements

Quantitative measurements of UV energy are possible
with the use of Black-Ray UV meters. These meters read
in $\mu W \ cm^{-2}$ and have no response to visible light. They
are available either as a short-wave, or a long-wave
UV meter from Ultra-Violet Products, Inc. 5100 Walnut
Grove Ave., San Gabriel, Ca. 91778 U.S.A.

b. Full-spectrum measurements

1) The ISCO Model SR spectroradiometer gives continuous scanning of wavelengths from 380 to 1500 nm.
Measurements are made in $\mu W \ cm^{-2} nm^{-1}$. This instrument
has been found to give accurate measurements both in
the laboratory and in the field. Its advantages include: 1) being very compact and portable, 2) works
on both A.C. and D.C., 3) capable of readings from
0.1 $\mu W \ cm^{-2}$ to 10,000 $\mu W \ cm^{-2}$ (bright sunlight), 4)
has attachments for a remote probe and a programmed
continuous scanning recorder.

The instrument is available from ISCO, Lincoln,
Nebraska 68504, U.S.A.

2) The Model 740 Optical Radiation Measurement System
capable of making accurate spectroradiometric, radio-
metric, and photometric measurements. Accessories extend the measurement capability to include spectral
radiance and luminance. The calibrated wavelength
range is from 300 to 1050 nm.

This instrument is available from Optronic Laborato-
ries, Inc., 7676 Fenton St., Silver Spring, Md. 20910
U.S.A.

G. Effect of Osmotic Potential on *Phytophthora*

The optimum growth of *P. cinnamomi* was obtained at osmotic potentials between -5 and -15 bars. Growth was reduced between
-20 and -25 bars, and completely inhibited between -40 and -50
bars, (Sommers *et al.* Phytopathology 60: 932-934 (1970), and
Adebayo, A.A., and R.F. Harris, Proc. Soil Sci. Am. 35: 465-
469 (1971).

Sterne *et al.* (Phytopathology 66: 1398-1402, 1976), have shown
that there is an interrelationship between specific ions and
osmotic potential in the inhibition of growth of *P. cinnamomi*.
When NaCl or KCl concentrations were adjusted to give an osmoti
potential of -9 bars, the fungus showed a 50% growth reduction.
Equivalent growth reductions were achieved with $CaCl_2$ and $MgCl_2$
at -11 and -16 bars respectively. With $MgSO_4$, a 50% growth reduction of the fungus was observed at -4 bars.

The influence of matric potential on the production of sporang
has been demonstrated by Sneh and McIntosh (1974), Duniway
(1975), and Pfender *et al.* (1977).

The maximum sporangia production of *P. megasperma* was obtained
in flooded soil, decreasing in soil at -0.05 and -0.1 bar
matric potential. No sporangia were produced at -3.0 bars
(Pfender *et al.*, Phytopathology 67: 657-663, 1977). Similar
results were obtained with *P. cactorum* by Sneh and McIntosh
(Can. J. Botany, 57: 795-802, 1974). In contrast, Duniway
(Phytopathology, 65: 1089-1093, 1975), reported that numerous
sporangia of *P. drechsleri* formed at -2.1 to -3.5 bars, but fe
to none were produced in flooded soil.

There is also apparently a temperature/soil moisture interacti
involved in the formation of sporangia, e.g. with *P. cactorum*,
over two times as many sporangia were produced at -0.1 bar tha
at -0.3 bar at $15^{O}C$. At $29^{O}C$, however, there was only a slight
difference in sporangia production between -0.1 bar and -0.3
bar (Sneh and McIntosh, 1974).

Over 70% germination of *P. cinnamomi* chlamydospores was obtain

at zero to -0.10 bar matric potential, but only 40-50% germination was obtained at -0.25 bar. Amendment of the soil with
glucose and asparagine at -0.25 bar, resulted in chlamydospore
germination equivalent to that obtained in non-treated soil at
zero to -0.10 bar matric potential (Sterne et al. Phytopathology
57: 1495-1500, 1977). It has also been demonstrated that the
serverity of root disease caused by chlamydospores of P. cinnamomi was a function of matric, but not of osmotic potential.
The percentage of diseased roots averaged from 80-90% at zero
matric potential, to 10-15% at -0.10 bar matric potential. Only
a few lesions were observed on roots of seedlings at -0.25 bar.

Useful reference:

COOK, R.J. and R.I. PAPENDICK. 1972. Influence of water potential of soils
and plants on root disease. Annual Review Phytopathol. 10: 349-374.

K. Protoplasts

Protoplasts of Phytophthora have been successfully established.
Details of the technique involved can be found in the following
publications.

BARTNICKI-GARCIA, S. and E. LIPPMAN. 1966. Liberation of protoplasts from the mycelium of Phytophthora. J. Gen. Microbiol. 42:
411-416.

BARTNICKI-GARCIA, S. and E. LIPPMAN. 1967. Enzymatic digestion
and glucan structure of hyphal walls of Phytophthora cinnamomi.

L. Biochemical Aspects

1. Cytochromes

The following absorption peaks for cytochromes a, b, and
c have been reported in Phytophthora capsici mycelium.

a-type.....cytochromes exhibit a broad peak with absorption maxima between 592-597 nm.

b-type.....cytochromes have absorption maxima at 557 nm
and 564 nm.

c-type.....cytochromes have absorption maxima at 543 nm
552 nm and 556 nm.

Note:

Only cytochrome c has been detected in oospores.

RIBEIRO, O.K., G.A. ZENTMYER, and D.C. ERWIN. 1976. Phytopathology 66: 172-
174.

In other Oomycetes, a, b, and c cytochromes have absorption
maxima at 605, 564, and 551 nm, respectively.

2. DNA content of some *Phytophthora* spp.

The base composition of deoxyribonucleic acid (DNA) of
some *Phytophthora* spp. is given in Table 13.

Table 13: Base composition of deoxyribonucleic acid (DNA) of some
Phytophthora spp.

Species	% GC content
P. boehmeriae	52.5
P. cactorum	53.5
P. colocasiae	58.0
P. cinnamomi	49-57
P. cryptogea	52
P. drechsleri	50
P. fragariae	54
P. heveae	55
P. infestans	54
P. nicotianae var. parasitica	50.5
P. nicotianae var. parasitica	49.0
P. palmivora	53.0

from F. GLEASON. Physiology of the lower fresh water fungi. *In*, Recent advances in aquatic mycology. E.B. Gareth-Jones (ed.). Reproduced by permission of John Wiley & Sons, Inc. Copyright, 1976.

See also: STORCK, R. and C.J. ALEXOPOULOS. 1970. Deoxyribonucleic Acid of
Fungi. Bacteriol. Rev. 34: 126-154.

3. Ribosomes

Vegetative mycelium of several *Phytophthora* spp. contain 80s ribosomes with subunits of 60s and 40s. The ribosomal RNA is similar to known RNA's of 28s and 18s values.

In dormant oospores of *Phytophthora*, there is apparently a lack of typical intact functional 80s ribosomes. However, subribosomal ribonucleoprotein particles, ribosomal RNA and ribosomal protein were isolated from oospores.

EARY, J.V., J.R. ROHEIM, and G.A. ZENTMYER. 1974. Phytopathology 64: 404-08.

4. Polar lipids

P. nicotianae var. *parasitica* is reported to differ significantly in polar lipid composition compared to other fungi.

HENDRIX, J.. and G. ROUSER. 1976. Mycologia 68: 354-361.

5. Chloramphenicol sensitivity

It has been shown that the site of action of chloramphenicol sensitivity in *P. palmivora* is the mitochondrion, and the specific site is believed to be the mitochondrial ribosome.

ARNES, D.W. 1976. The sensitivity of *Phytophthora palmivora* to chloramphenicol. Ph.D. Dissertation. Univ. of Calif., Riverside, Ca. 75 p.

Section 16: Control

16A Chemical control using fungicides

I. The control of *Phytophthora* in soil presents many problems
    due to:

    1. Its ability to produce secondary inoculum from primary
        survival structures.

    2. Its ability to invade soil at considerable depths unlike
        most root infecting fungi. It can thus readily colonize
        living tissue that has been plowed under.

    3. Its ability to penetrate soil deeply allows it to escape
        the majority of antagonists which do not thrive at these
        depths.

    4. It is not economically feasible to use standard soil
        fumigants due to the depth at which *Phytophthora* can
        exist in the soil.

    5. Its ability to produce several different forms of inocu-
        lum which have varying degrees of resistance to environ-
        mental conditions, e.g. mycelium, sporangia, zoospores,
        cysts, chlamydospores, and oospores.

    6. Its wide host range, e.g. *P. cinnamomi* parasitizes over
        900 different plants; *P. palmivora* attacks over 51 genera
        of plants; (see Hosts Section, pages 10-15, for details of
        hosts of other species of *Phytophthora*).

    7. Zoospores can concentrate near host root tips within a
        few hours after zoospores release from the sporangium
        (see Chemotaxis and Electrotaxis).

    8. *Phytophthora* can grow at oxygen levels injurious to most
        roots (Stolzy *et al.*, Phytopathology 55: 270-275, 1965).

    9. The rapid spread of the disease in poorly drained or
        water-logged soils.

I. General practices advocated for control of *Phytophthora*
    diseases

    1. Good soil drainage. This is probably the single most im-
        portant factor for a majority of *Phytophthora* diseases.

    2. Disease-free seed.

3. Crop rotation

4. Deep ploughing.

5. Resistant varieties and/or bud trees on resistant root-
   stock at least 6 in. above soil line.

6. Fallowing

7. Fungicides.

8. Fumigation of nursery soils, and tree planting sites.

III. Some measures advocated for specific *Phytophthora* diseases

a) *P. cinnamomi*

Use clean seed or material treated in water between 49-54°C
for 30 min. Plant in fumigated soil in a well-drained area.
Do not over-irrigate. In tropical rain forest soils, the
following measures are advocated:

a) plant in deep well-drained soils, avoid water-logging;
b) use resistant rootstocks; c) maintain high soil calcium
levels (don't exceed 1 ton/ha per application; d) maintain
high soil nitrogen levels, but prevent rapid conversion to
nitrate nitrogen; e) avoid clean cultivation; grow cover
crops such as forage sorghum or corn and a legume crop in
summer, and follow up with blue lupin (*Lupinus angustifolius* L.) as a winter crop. Disc crops into soil when mature
Mulch large trees with low nitrogen containing materials
such as wheat straw or sorghum stubble (Pegg, K.G., Calif.
Avocado Soc. Yearbook, 1976).

b) *P. citrophthora*

Use resistant stock not budded too close to the soil. Choos
well-drained sites. Avoid injury to trunks of trees. Clear
trash and weeds from the base of the tree.

c) *P. infestans*

Destroy volunteer growth on cull piles or other tuber re-
fuse. Use resistant varieties.

d) *P. palmivora*

An extensive review of control methods is presented in "*Phy-tophthora* disease of Cocoa," P.H. Gregory (ed.), Longman Grp. Ltd. London (see pages 211-268).

e) *P. phaseoli*

Zinebs such as Parzate and Dithane Z78 are reported to be effective. Control insect population visiting diseased blossoms. (See Sturgis, Botanical Gazette 25: 187-194 (1898) for illustrations of the mechanisms involved in the dissemination of sporangia of *P. phaseoli* by bees.)

IV. Chemicals reported to be effective against various *Phytoph-thora* species

1. Dexon (p-dimethylaminobenzenediazo sodium sulfonate)-Chemagro Corp.

2. Terrazole (5-ethoxy-3-trichloromethyl-1,2,4-thiadiazole)-Olin Chemical Co. Also known as Truban, Ethazole, etc.

3. Pyroxychlor (2-chloro-6-methoxy-4 [trichoromethyl] pyridine- Dow Chemical Co.

4. Dowco 269.

5. Bordeaux mix - toxic to the foliage of some plants, e.g. tomato.

6. Copper fungicides, e.g. copper oxychloride, copper oxide, etc.

7. Dithiocarbamates, e.g. captan, zineb.

a) Dexon

Dexon (Chemoagro Chemical Corp.) is reported to have controlled *Phytophthora* root rot of avocado (*Persea americana* Mill.) seed-ings in glasshouse tests. However, repeated applications were necessary. Some control is also obtained with large avocado

Table 14: List of some *Phytophthora* diseases partially or completely controlled by specific fungicides

Chemical	Pathogen	Disease
Dexon	*P. cinnamomi*	Avocado Root Rot
Terrazole	*P. cinnamomi*	Avocado Root Rot
Bordeaux mix	*P. infestans*	Late Blight of Potatoes
Copper oxychloride	*P. infestans*	Late Blight of Potatoes
Copper oxide	*P. infestans*	Late Blight of Potatoes
Dithiocarbamates	*P. infestans*	Late Blight of Potatoes
Pyroxychlor	*P. megasperma* var. *sojae*	Root Rot of Soybeans
Dichlone	*P. nicotianae* var. *parasitica*	Black Shank of Tobacco
Captafol	*P. nicotianae* var. *parasitca*	Black Shank of Tobacco
Pyroxychlor	*P. nicotianae* var. *parasitica*	Black Shank of Tobacco
Terrazole	*P. nicotianae* var. *parasitica*	Black Shank of Tobacco
Lorvek	*P. nicotianae* var. *parasitica*	Root Rot of *Pinus radiata*
Dowco 269	*P. nicotianae* var. *parasitica*	Root Rot of *Pinus radiata*
Copper oxychloride	*P. palmivora*	Black pod of Cocoa
Copper hydroxide	*P. palmivora*	Black pod of Cocoa
Ortho-difolatan	*P. palmivora*	Black pod of Cocoa
Bordeaux mix	*P. palmivora*	Black pod of Cocoa
Triphenyltin hydroxide	*P. palmivora*	Black pod of Cocoa
Triphenyltin acetate	*P. palmivora*	Black pod of Cocoa
Triphenyltin chloride	*P. palmivora*	Black pod of Cocoa
Pyroxychlor	*P. palmivora*	Root Rot of *Dieffenbachia*
Pyroxychlor	*P. palmivora*	Root Rot of *Rhododendron*

trees in the field with continued applications of Dexon.

The action of Dexon is fungistatic: 5-20 ppm in the soil will
reduce formation of sporangia and chlamydospores by 90-100%.

Disadvantages

1. Lack of persistence in soil: repeated applications are required for control.

2. Sensitivity to light (apply under reduced light conditions and water chemical immediately into soil.

ZENTMYER, G.A. 1973. Phytopahtology 63: 267-272.

b) Pyroxychlor

Pyroxychlor, a systemic fungicide, has been used effectively in the control of some *Phytophthora* diseases. *In vitro* sensitivity (ED 50) of mycelial growth of 10 *Phytophthora* spp. ranged from 0.4 to 6.2 ppm. *Phytophthora* spp. tested were: *P. cactorum*, *P. capsici*, *P. cinnamomi*, *P. cryptogea*, *P. drechsleri*, *P. erythroseptica*, *P. fragariae*, *P. infestans*, *P. megasperma*, and *P. nicotianae* var. *parasitica*. (Richardson, L.T. Can. J. Plant Sci. 56: 365-369, 1975.)

c) Control of *P. megasperma* var. *sojae* by acetone infusion of pyroxychlor

1. Dissolve the systemic fungicide, Pyroxychlor (125 mg active ingredient/100 g seed), in pure acetone to give a 4% solution.

2. Soak soybean seeds in the 4% Pyroxychlor-acetone solution for 30 min. before planting.

3. Five weeks after planting, significantly higher stands of soybean seedlings are obtained by this method than by other methods, e.g. the slurry method using adherents such as gum arabic.

4. No phytotoxicity has been observed on plants using this acetone infusion method.

PAPAVIZAS, G.C. and J.A. LEWIS. 1976. Plant Disease Reptr. 60: 484-488.

d) Treatments for *Phytophthora* foot rot of citrus: *P. nico-
    tianae* var. *parasitica*

1. Captan and other copper fungicides at 60 mg/ml gave 100%
    inhibition of *Phytophthora*, while only 0.6 mg/ml of
    captafol was required to give the same effect. The
    amounts are based on active ingredient or metallic Cu.

2. Trunk paints at 60 mg/ml, indicate that copper ammonium
    carbonate, cupric hydroxide, and captafol are active for
    at least 33 weeks. Captan and pyroxychlor are active for
    less than 17 weeks.

3. A trunk paint of 2.5% captan + 2.5% copper fungicide in
    bentonite at a concentration of 0.48 g/ml remains active
    for at least 28 weeks.

TIMMER, L.W. 1977. Phytopathology 67: 1149-1154.

Note:

Losses caused by *P. nicotianae* var. *parasitica* are considerably
higher with citrus trees planted on level ground than on those
planted on raised ridges (Timmer and Leyden [1976], J. Rio
Grande Valley Hort. Soc. 30: 19-25).

V. Laboratory methods for testing soil fungicides

1. Sift air-dried fine sandy loam (pH 7 to 7.5) through a
    20 mesh sieve.

2. Place the sieved soil in 2 qt. mason jars, and autoclave
    for 45 min. at 15 lb/sq. in.

a) Evaluation of fungicides applied in solution or suspension

1. Place 1 in. of this soil in a shell vial of 25 cc capaci
    ty (20 mm in diameter x 85 mm deep).

2. Cut a 10 mm diameter disc from the outer margin of a potato - dextrose agar culture of the fungus to be tested, and place on the soil surface.

3. Cover the disc with 1 in. of steam sterilized soil.

4. Pipette 5 ml of the fungicidal solution or suspension to be tested on to the surface of the soil. This is sufficient to wet all of the soil in the vial. Make three replicate vials.

5. Incubate vials for 24 hrs at 25°C.

6. After 24 hrs, empty the vials into a wire or perforated metal strainer, and remove the soil by washing with running water.

7. Pick up the discs with sterile forceps, blot on paper towels, and place on cornmeal agar (see page 76)   in petri dishes to determine viability of the fungus.

8. Cornmeal agar is a good medium for the growth of *Phytophthora*. It also has the advantage of sufficiently limiting the growth of contaminants so that viability of the test fungus can easily be ascertained.

b) Evaluation of fungicides mixed dry with the soil

1. Place 60 g of steam-sterilized soil in a 200 ml Erlenmeyer flask.

2. Add a measured amount of the dry fungicide to the flask.

3. Mix soil and fungicide thoroughly by shaking.

4. Place 1 in. of the treated soil in the bottom of each of 3 vials.

5. Add the inoculum disc and cover with 1 in. of treated soil.

6. Add 5 ml of distilled water to the surface of the soil.

7. Incubate for 24 hrs at 25°C, and recover the discs by the method described in I (7-8) above.

316

CONTROL

Notes:

a) To accurately obtain low dosages (1-5 ppm), add the fungicide to the soil at a higher concentration (25 ppm), and mix thoroughly. Then add soil to the mixture until the desired concentration is reached.

b) There is usually little correlation between fungicide tests conducted in agar plates, and in soil tests.

c) The following factors influence results obtained in fungicide tests in soil: i) pH soil type, ii) age of inoculum (1-day-old and 8-day-old mycelium react differently to the chemical, iii) inoculum potential (a lower dosage is required to kill a fungus when a 5 mm disc of inoculum is used, than when a 10 mm disc is used.

ZENTMYER, G.A. 1955. Phytopathology 45: 398-404.

See also, McINTOSH, D.L. 1971. Plant Disease Reptr. 55: 213-216.

c) Laboratory bioassay for sprayed plant material

1. Cut discs from leaves 3 hrs, 5 days and 13 days after spraying.

2. Place on agar plates and measure inhibition zones formed after inoculating the agar plates with a *Helminthosporium* sp.

CHAVES, G.M. 1954. Turrialba 4: 39-42.

Section 16: Control

16B Biological control

The following methods of biological control have been reported.

I. Mycorrhizal fungi

P. cinnamomi and some other *Phytophthora* spp. are inhibited by
volatile monoterpenes and sesquiterpenes from certain mycorrhizal fungi. It has been shown that increasing the abundance of
mycorrhizal fungi on *Pinus radiata* roots can protect against
infection by *Phytophthora*.

KRUPA, S. and J.E. NYLAND. 1972. European J. Forest Pathol. 2: 88-94.

MARX, D.H. 1975. Pages 112-115 in "Biology and Control of Soil-borne Plant
Pathogens," G.W. Bruel (ed.), Amer. Phytopath. Soc., St. Paul, Minn., U.S.A.

II. Suppressive soils

Some natural soils are reported to suppress growth and reproduction of *Phytophthora*. Several factors appear to be involved.
These include content of organic matter, high calcium levels,
favorable pH, high levels of ammonium and nitrate nitrogen,
adequate phosphate, high population of microorganisms, and high
magnesium content. Some suppressive soils have been shown to
contain antagonists that 1) lyse hyphae and sporangia, 2) contain inhibitory agents that inactivate sporangia production and,
3) have good drainage. *Phytophthora* suppression is overcome by
waterlogging, addition of large amounts of inoculum, or sub-
terranean clover meal.

BROADBENT, P. and K. BAKER. 1974. Austral. J. Agr. Res. 25: 121-137.

BAKER, K.F. and R.J. COOK. 1974. Biological Control of Plant Pathogens.
Freeman and Son. San Francisco, 433 p.

Ko (Phytopathology 61: 780-782, 1971), has reported successful
biological control of root rot of papaya seedlings caused by *P.
almivora*. Planting holes were dug in *Phytophthora* infested
field soil and filled with pathogen-free virgin soil. Healthy

papaya seedlings were then planted in these holes. By the time
the roots spread to the surrounding infested soil, the seedling
had become sufficiently resistant to P. *palmivora* so that littl
disease developed. Seedlings had thus been protected at the mos
susceptible stage.

III. Soil amendments

The following soil amendments have been reported to reduce the
incidence of avocado root rot by P. *cinnamomi*.

    a) Alfalfa meal amended soil.

    b) Cotton waste.

    c) Soybean meal.

    d) Wheat straw.

Caution:

Alfalfa meal and soybean meal have both been reported to be
phytotoxic under certain conditions.

Materials reported to be effective against P. *cinnamomi* and P.
nicotianae var. *parasitica* in laboratory tests include:

Chicken manure (2 or 4%, w/w)

Alfalfa meal (2 or 4%)

Hydrolyzed feathermeal (1-2%)

Urea (0.1 or 0.2%).

Caution:

Some amendments used at very high concentrations can be phytotoxic.

GILPATRICK, J.D. 1969. Phytopathology 59: 979-985.

SAO, P.H. and G.A. ZENTMYER. 1977. Phytophthora Newsletter 5: 48.

ZENTMYER, G.A. 1963. Phytopathology 53: 1383-1387.

V. Weed control

Milkweed vine, *Morrenia ordorata* Lind., a major weed pest in
Florida citrus groves can be biologically controlled by a
pathotype of *P. citrophthora*. This isolate of *P. citrophthora*
has thus far been shown to be stable, although other isolates
of *P. citrophthora* are known to be pathogenic to citrus.

RIDINGS, W.H. *et al*. 1976. Proc. Amer. Phytopathol. Soc. 3: 273-274.

V. Tests for antagonistic microorganisms

See method described by Broadbent *et al*. Austral. J. Biol. Sci.
48: 237-248.

Section 17:

Origins of *Phytophthora*

I. Origin of *P. cinnamomi*

P. cinnamomi is not native to soils in California or Latin America (Zentmyer, 1976, & 1977). It is also believed not to be native to Australia and New Zealand (Newhook and Podger, 1972), but introduced to these regions in the 18th century. However, Shepherd (1975) believes that the fungus is native to eastern Australia, having entered over a million years ago from the New Guinea-Celebes region. Zentmyer (1976 & 1977) postulates that *P. cinnamomi* was first carried to the western hemisphere with plant material and accompanying soil from the Malaysian-eastern Australian area on early voyages and explorations throughout the Pacific and the Americas. He believes that the fungus was brought into California in late 19th and early 20th centuries with the introduction of avocados and other tropical plants from Hawaii and Central America.

NEWHOOK, F.J. and F.D. PODGER. 1972. Ann. Rev. Phytopath. 10: 299-326.

SHEPHERD, C.J. 1975. Search 6: 484-490.

ZENTMYER, G.A. 1976. California Avocado Soc. Yearbook 60: 154-158.

ZENTMYER, G.A. 1977. Phytopathology 67: 1373-1377.

II. Origin of *P. infestans*

Reddick (1939) believes that the original introduction of the pathogen to the Old World was possibly by oospores from Solanaceous plants introduced for ornamental or pharmaceutical purposes. The origin of the fungus appears to be in Mexico since it occurs on wild native species. Most of these species are resistant or immune; cultivated potatoes are rare. Origins of *P. infestans* in South America, e.g. Peru or Chile, can be discounted due to a lack of resistant varieties. Also, the fungus is present only where European potatoes are grown.

The narrow host range of *P. infestans* is believed to be a consequence of its origin in an area of comparatively little cultivation and restricted climate, unlike *P. palmivora* (see below).

III. Origins of *P. palmivora*

Waterhouse (1977) speculates that the possible center of origin
of *P. palmivora* may lie either in the Indo-Pacific, or in Central or South America, since a majority of the reported hosts
are natives of either Asia or Central and South America. These
apparently widely separated regions were once closely associate
as the southern and tropical parts of the ancient continent of
Gondwanaland, their present position being the result of conti-
nental drift.

According to Waterhouse (1977) *P. palmivora* most likely evolved
from a completely aquatic or wetland form to be semi-aerial,
the transition probably being made in a host with leaves or
fruits touching or close to the soil. This would more readily
occur in a damp climate.

If Central or South America is given consideration as the origins of *P. palmivora*, excellent candidates as native hosts
would be wild cocoa and rubber. Both these trees grow in very
wet spots. The spread of *P. palmivora* to the Old World would
then have occurred with the transport of cocoa via Central
America and the West Indies in the 16th century. On the other
hand, there are numerous endemic hosts in the Indo-Pacific (e.
g. coconut palm, bread-fruit, etc.), if one considers this region as the origin of the pathogen. Transport to the New World
would be either west via the West Indies, or east across the
Pacific Islands to Columbia, Ecuador, and Peru. If given a
choice, it seems plausible that the Old World crops and cultiva
tion practices offered a better chance for the development and
spread of *P. palmivora*.

WATERHOUSE, GRACE M. 1977. Phytophthora Newslatter 5: 3-5.

Section 18:

Some aids for the identification

of *Phytophthora* species

I. Taxonomy

It is difficult to apply a natural system of classification to
the genus *Phytophthora* with the data currently at our disposal.
The present arbitrary methods of classification are based on
morphological characteristics which tend to overlap species
descriptions, creating confusion in identification and classification.

The biggest problem facing the taxonomist is whether a particular pathotype possesses sufficiently stable new characters to
warrant a *species novo* or a sub-species classification. This
dilemma will only be resolved when the extent of biological
variation within each species is known. A promising beginning
in this direction is presently being undertaken at the University of California, Riverside. Hundreds of isolates of *P. cinnamomi, P. megasperma,* and *P. palmivora,* collected world-wide
from several different hosts, are being carefully studied to
delimit the extent of variation between isolates of the same
species for several morphological and physiological characteristics. This is obviously a monumental task and by its very nature time-consuming.

The problem of taxonomy is further compounded by the fact that
species inter-cross readily, producing viable oospores. Are
some of the described species with close affinities to other
species in reality inter-specific hybrids? One is also faced
with the perplexity of trying to classify isolates that meet
the criteria of a described species in most respects, but are
most specific.

It appears that there are, in all probability, only a dozen or
so valid species of *Phytophthora;* the rest are extensions of
variation attributable to adaptation, hybridization between
species, and/or physiologic specialization. This radical view
will have to await the accumulation of data indicating the close
affinity of many of the described species. It is of interest to
note that over the years, several *Phytophthora* species have been
combined in the light of recent studies, e.g. *P. fagi, P. omni-
ora, P. paeoniae,* are now combined into *P. cactorum; P. faberi,*

and *P. theobromae,* are now considered to be the same as *P. pal-*
mivora; P. drechsleri and *P. cryptogea* are considered to be
identical. Also, recent serological studies indicate that *P.*
arecae, P. citrophthora, P. meadii, P. mexicana, and *P. palmi-*
vora are identical and should be combined into one species -
P. citrophthora.

Useful references:

1. BLACKWELL, E. 1949. Terminology in *Phytophthora.* Mycol. Papers No. 30:
   25 p. Commonwealth Mycol. Instit. Kew, Surrey, England.

2. FREZZI, M.J. 1950. Las especies de *Phytophthora* en la Argentina. Revista de Invest. Agric. t-iv. 1: 47-134. B. Aires, Argentina.

3. LEONIAN, L.M. and L.H. GREER. 1929. Comparative value of the size of
   Phytophthora sporangia obtained under standard conditions. J. Agric.
   Res. U.S.A. 39: 293-311.

4. LEONIAN, L.M. and L.H. GREER. 1934. Identification of *Phytophthora*
   species. Bull. Agric. Expt. Sta. W. Va. Univ. 262: 136.

5. MERZ, W.G., R.G. BURRELL, and M.E. GALLEGLY. 1969. A serological comparison of six heterothallic species of *Phytophthora.* Phytopathology
   59. 367-370.

6. NEWHOOK, F.J., G.M. WATERHOUSE, and D.J. STAMPS. 1978. Tabular key to
   the genus *Phytophthora.* Mycol. Paper. Commonwealth Mycol. Instit. Kew,
   Surrey, England.

7. SAVAGE, E.J. *et al.* 1968. Homothallism, heterothallism, and interspecif
   ic hybridization in the genus *Phytophthora.* Phytopathology 58: 1004-10᷈

8. TUCKER, C.M. 1931. Taxonomy of the genus *Phytophthora.* Missouri Agr.
   Expt. Sta. Bull. 153: 208 p.

9. WATERHOUSE, G.M. 1963. Key to the species of *Phytophthora* de Bary. Commonwealth Mycol. Instit. Kew, Surrey, England. Mycol. Paper No. 92: 22

10. WATERHOUSE, GRACE M. 1970. The genus *Phytophthora* de Bary. Mycol. Pape᷈
    No. 122. Commonwealth Mycol. Instit. Kew, Surrey, England.

11. WILSON, G.W. 1914. Studies of North American Peronosporales: V. Review
    of the genus *Phytophthora.* Mycologia 6: 54-83.

XI. Table 15. List of currently accepted *Phytophthora* species

Species	Year	Authority
1. *P. arecae*	1913	(Coleman) Pethybridge
2. *P. bahamensis*	1975	Fell & Master
3. *P. boehmeriae*	1927	Sawada
4. *P. botryosa*	1969	Chee
5. *P. cactorum*	1886	(Lebert & Cohn) Schroeter
6. *P. cambivora*	1927	(Petri) Buisman
7. *P. capsici*	1922	Leonian
8. *P. carica* (fici)	1915	(Hara) Hori
9. *P. castaneae*	1968	Katsura and Uchida
10. *P. cinnamomi*	1922	Rands
11. *P. citricola*	1927	Sawada
12. *P. citrophthora*	1925	(Smith & Smith) Leonian
13. *P. colocasiae*	1900	Raciborski
14. *P. cryptogea*	1919	Pethybridge and Lafferty
15. *P. cyperi*	1935	(Ideta) Ito
16. *P. cyperi-bulbosi*	1953	Seethalakashmi & Ramakrishnan
17. *P. drechsleri*	1931	Tucker
18. *P. epistomium*	1975	Fell and Master
19. *P. erythroseptica*	1913	Pethybridge
20. *P. fragariae*	1940	Hickman
21. *P. gonapodyides*	1909	Petersen
22. *P. heveae*	1929	Thompson
23. *P. hibernalis*	1925	Carne
24. *P. infestans*	1876	(Montagne) de Bary
25. *P. inflata*	1949	Caroselli and Tucker
26. *P. lateralis*	1942	Tucker and Milbrath
27. *P. lepironiae*	1919	Sawada
28. *P. macrospora*	1940	(Saccardo) Ito and Tanaka
29. *P. meadii*	1918	McRae
30. *P. megasperma* var. megasperma	1931 1963	(Drechsler) Waterhouse
31. *P. megasperma* var. sojae	1959	(Drechsler) Hildebrand
32. *P. melonis*	1967	Katsura
33. *P. mexicana*	1923	Hotson and Hartge
34. *P. mycoparasitica*	1975	Fell and Master
35. *P. nicotianae* var. nicotianae	1896 1963	(Breda de Haarn) Waterhouse

Table 15: List of *Phytophthora* species (continued)

Species		Year	Authority
36.	*P. nicotianae* var. *parasitica*	1896 1963	(Dastur) Waterhouse
37.	*P. oryzae*	1931	(Ito and Nagai) Waterhouse
38.	*P. palmivora*	1919	(Butler) Butler
39.	*P. phaseoli*	1889	Thaxter
40.	*P. porri*	1931	Foister
41.	*P. primulae*	1952	Tomlinson
42.	*P. quininea*	1947	Crandall
43.	*P. richardiae*	1927	Buisman
44.	*P. spinosa* var. *lobata*	1975	Fell and Master
45.	*P. spinosa* var. *spinosa*	1975	Fell and Master
46.	*P. syringae*	1909	(Klebahn) Klebahn
47.	*P. verrucosa*	1940	Alcock and Foister
48.	*P. vesicula*	1969	Anastasiou and Churchland
49.	*P. vignae*	1957	Purss

III. Caducity

Sporangia of some species of *Phytophthora* separate readily from the sporangiophore. This property of the sporangium has been termed "caducity". The sporangium detaches from the sporangiophore with a specific pedicel length that is characteristic for the species, (see Table 16).

Not all *Phytophthora* species are caducous. Species that inhabit above ground plant parts, e.g. *P. infestans* and *P. phaseoli*, are normally caducous. These species are all papillate with deep apical thickening. Noncaducous or persistent sporangia belong to species that usually attack the roots and/or collars of plants. These sporangia are non-papillate with shallow apical thickenings.

Sporangia are rarely detached before maturity has been attained. Other factors that affect sporangium caducity are temperature, duration of exposure to light and dark periods, and the source

ɔf the isolate (species isolated on   aerial parts of the plant
are usually caducous).

Sporangium caducity,   pedicel length, and ontogeny are stable
diagnostic characters in *Phytophthora* and Tsao (1977) suggests
that descriptions of new and existing species   should include
these diagnostic characteristics.

A list of caducous and non-caducous species are given in Tab. 16.

able 16: Caducous and persistent species of *Phytophthora*

Caducous		Persistent (not shed)	
hed in dry air nd in water	Shed in water only	Papillate	Non-papillate
nfestans (< 3 μm)☆	*arecae* (1-6 μm)	*citricola*	*cambivora*
haseoli (short)	*boehmeriae* (< 3 μm)	*inflata*	*cinnamomi*
	botryosa (short)	*mexicana*	*cryptogea*
	cactorum (< 4 μm)	?*nicotianae*	*drechsleri*
	capsici (10 μm+)	*palmivora*	*erythroseptica*
	carica (10-20 μm)	pepper form	and var. *pisi*
	citrophthora?	*porri*	*fragariae*
	colocasiae (3.5-10 μm)	*primulae*	*gonapodyides*
	cyperi (4 μm)	*syringae*	*lateralis*
	cyperi-bulbosi?		*megasperma* and
	fici (short)		var. *sojae*
	heveae (10 μm)	*oryzae*	*oryzae*
	hibernalis (55 μm)		*quininea*
	ilicis (5-15 μm)		*richardiae*
	lepironiae (short)		*verrucosa*
	meadii (10-15 μm)		*vesicula*
	nicotianae var.		*vignae*
	parasitica (2 μm)		
	palmivora		
	form 1 (2-10 μm)		
	form 2?		

Length of stalk.

rom WATERHOUSE, G.M. *In* "*Phytophthora* disease of Cocoa", (P.H. Gregory,
d.). By permission of Longman Grp. Inc. Copyright, 1974.

Note:

Most caducous species shed their sporangia in water. Sporangia of some species, e.g. *P. infestans* and *P. phaseoli*, can also be detached by air currents (Waterhouse, 1974).

IIIa. L:B ratios

The ratio of the length of the sporangium in relation to its breadth (L:B or l:b ratios), have often been used in describing the morphology of sporangia of *Phytophthora* species. This characteristic is now known to vary considerably depending on the substrate and the quantity and quality of light employed (see Hendrix, Mycologia 59: 1107-1111, 1967 and Zentmyer and Ribeiro, Phytopathology 67: 91-95, 1977).

L:B ratios are therefore of little diagnostic value unless precise standard conditions are stipulated.

IV. Grouping of *Phytophthora* species according to Waterhouse (1963)

Group I

P. cactorum	Apex of sporangium markedly papillate. Antheridia paragynous.

Grouping of *Phytophthora* (continued)

Group II

P. *arecae*

P. *boehmeriae*

P. *botryosa*

P. *capsici*

P. *carica (fici)*

P. *castaneae*

P. *citrophthora*   Apex of sporangium markedly

P. *heveae*   papillate. Antheridia amphi-

P. *meadii*   gynous. See Morphology section.

P. *melonis*

P. *mexicana*

P. *nicotianae* var.
     nicotianae

P. *nicotianae* var.
     parasitica

P. *palmivora*

P. *vesicula*

Group III

P. *citricola*

P. *cyperi*

P. *cyperi-bulbosi*

P. *inflata*   Apex of sporangium not as

P. *lepironiae*   protuberant as Groups I and II.

P. *macrospora*   Sometimes scarcely papillate.

P. *porri*   Oospores always present.

P. *primulae*   Antheridia predominantly

P. *syringae*   paragynous. See Morpohology

                           Section.

Grouping of *Phytophthora* (continued)

Group IV

P. *colocasiae*

P. *hibernalis*   Apex of sporangium as in Group

P. *infestans*   III. Antheridia predominantly

P. *phaseoli*   amphigynous. See Morphology
                          Section.

Group V

P. *fragariae*

P. *megasperma* var.   Sporangia non-papillate. Very
    megasperma
                           thin apical thickening.   See
P. *megasperma* var. *sojae*
                           Morphology Section.
P. *quininea*

P. *verrucosa*

Group VI

P. *cambivora*

P. *cinnamomi*

P. *cryptogea*

P. *drechsleri*   Sporangium non-papillate, pro-

P. *erythroseptica*   liferating internally. Antheri

P. *gonapodyides*   dia predominantly amphigynous.

P. *lateralis*   See Morphology Section.

P. *oryzae*

P. *richardiae*

P. *vignae*

V. Table 17: List of *Phytophthora* species grouped according to sexual type
and antheridial configuration

Homothallic Species		Heterothallic Species
Paragynous	Amphigynous	Amphigynous
P. *cactorum*	P. *cambivora*	P. *arecae*
P. *citricola*	☆P. *erythroseptica*	P. *botryosa*
P. *cyperi*	P. *fragariae*	P. *boehmeriae*
P. *cyperi-bulbosi*	P. *heveae*	P. *cambivora*
P. *fragariae*	P. *hibernalis*	P. *capsici*
P. *inflata*	P. *melonis*	P. *cinnamomi*
P. *lateralis*	P. *palmivora*	P. *citrophthora*
P. *lepironiae*	P. *phaseoli*	P. *colocasiae*
P. *macrospora*	P. *vignae*	P. *cryptogea*
☆P. *megasperma* var. *megasperma*		P. *drechsleri*
		P. *infestans*
☆P. *megasperma* var. *sojae*		P. *meadii*
P. *porri*		P. *mexicana*
P. *primulae*		P. *nicotianae* var. *nicotianae*
P. *quininea*		P. *nicotianae* var. *parasitica*
☆P. *richardiae*		
P. *syringae*		P. *palmivora*
☆P. *verrucosa*		
P. *vesicula*		

☆Species predominantly of this type. Sexual bodies are unknown in P. *carica* and P. *gonapodyides*.

VI. Table 18: Colony characteristics of *Phytophthora* species

Species	Colony characteristics✩
P. arecae	fairly fluffy, slightly radiate
P. bahamensis	rosette, superimposed with concentric growth rings
P. boehmeriae	uniform with dense aerial mycelium
P. botryosa	slightly radiate zonate, little aerial mycelium
P. cactorum	slightly fluffy and slightly radiate
P. cambivora	moderately to profusely fluffy and fairly uniform
P. capsici	finely radiate, fluffy
P. cinnamomi	profuse tough aerial mycelium; sometimes appressed
P. citricola	"chrysanthemum" pattern, fairly fluffy
P. citrophthora	characteristic finely radiate growth
P. colocasiae	uniformly fluffy
P. cryptogea	uniform and fairly fluffy
P. cyperi	known only on host
P. cyperi-bulbosi	known only on host
P. drechsleri	very fluffy, slightly rosette-like
P. erythroseptica	entirely non-fluffy
P. fragariae	fairly fluffy
P. gonapodyides	not known
P. heveae	very little aerial mycelium
P. hibernalis	rosette-like and fairly fluffy
P. infestans	fairly fluffy
P. lateralis	little or no aerial growth
P. macrospora	known only in host
P. meadii	variable aerial mycelium
P. megasperma var. *megasperma* var.	radiate and fluffy, some appressed
P. megasperma sojae	slightly radiate and fairly fluffy
P. mexicana	moderate aerial growth, surface 'frosty'
P. nicotianae var. *nicotianae*	diaphanously fluffy; no zones or striations
P. nicotianae var. *parasitica*	irregular rosette pattern
P. palmivora	uniform and slightly radiate
P. phaseoli	fairly fluffy
P. porri	radiate, aerial mycelium dense
P. quininea	cottony; cobweb-like aerial mycelium

Table 18: Colony characteristics of *Phytophthora* species (continued)

Species	Colony characteristics☆
P. *richardiae*	medium fluffy
P. *spinosa* var. *spinosa*	rosette, smooth
P. *syringae*	rose-shaped growth zones
P. *verrucosa*	not known in culture
P. *vesicula*	submerged aquatic fungus
P. *vignae*	some loose aerial mycelium

☆Based on cultures grown on corn meal agar medium. Colony characteristics
vary considerably on different media - see Ribeiro *et al.* Mycologia 67:
1012-1019 (1975).

WATERHOUSE, G.M. 1963. Mycol. Papers NO. 92: 22 p.

VII. Table 19: Separation of *Phytophthora* from *Pythium*

Characteristic	*Phytophthora*	*Pythium*
Sporangium shape	Usually ovoid or obpyriform. Never filamentous.	Filamentous, inflated hyphae, spherical. Occassionally ovoid or pyriform.
Sporangium position	Always terminal.	Terminal or intercalary.
Sporangium apical thickening (papillum)	+	-
Sporangium caducity	Caducous (deciduous) or persistent.	Always persistent
Zoospore differentiation	Protoplasm differentiates into zoospores within sporangium.	Protoplasm emerges undifferentiated from the sporangium into a spherical vesicle. Zoospores differentiate in the vesicle.
Hyphae	Coarse and stiffly branched (6–14 μm wide).	Fine and flexuous (5–10 μm).
Antheridia	Amphigynous or paragynous	Paragynous or hypogynous. Never amphigynous.
Oogonia	Smooth or occasionally ornamented. Pigmented walls.	Smooth or spiny. Colorless walls.
Oospore	aplerotic	plerotic or aplerotic
Chlamydospores and hyphal swellings	+	Rare
Thiamine requirement	+	- (except for *P. vexans* which requires thiamine for growth).
Habitat	Soil; aerial parts of plants; aquatic. Rarely saprophytic.	Soil or aquatic. Saprophytes or facultative parasites.
Addition of hymexazol (3-hydroxy-5-methylisoxazole) to isolation medium.	Not inhibited	Inhibited
Host specificity	Specific or numerous depending on species.	Numerous hosts, rarely host specific.
Number of species	49	92

Notes:
1. Protein and oxidase patterns are different for *Pythium* and *Phytophthora*.
2. The protein content of hyphal walls is greater in *Pythium*.
3. It is more difficult to release glucan-protein from *Pythium* than from *Phytophthora* cell walls.
4. Cell wall antigens from *Pythium* do not develop precipitin bands to any *Phytophthora* antisera.

VIII. Selection of sporangia for diagnostic purposes

It is important to use freshly developed sporangia for diagnostic purposes, since the apex in older sporangia can often
develop into a 'beak' or 'false papilla'.

Arrested development of the apex due to unfavorable conditions
prior to zoospore emission results in a 'false papillum'. The
characteristic 'beak' observed on some sporangia usually results from the cessation of germ tube development during direct
germination of sporangia. Unfavorable environmental conditions
are often responsible for this aberation.

WATERHOUSE, G.M. 1975. Phytophthora Newsletter 3: 5-7.

IX. Papilla

If the apical thickening is large in amount (3-5 μm or more
deep), it causes the apex of the sporangium to bulge outwards
to form a papillum. If the thickening is so shallow as to be
almost inconspicuous, then no papillum develops.

Sporangia with deep apical thickening usually have narrow exit
pores (av. 5 μm diam.), while those with inconspicuous thickening have wider exit pores (av. 12 μm diam.).

Stain with lactophenol/cotton blue, or rose bengal, since papilla are not marked by optical brighteners.

Note:

Diethanol fluorescence clearly differentiates the pedicel and
its attachment to the sporangium. It also differentiates the
point of emergence of the germ tube in the sporangium wall from
the rest of the sporangium structure.

GISI, U. and F.J. SCHWINN. 1976. Microscopica Acta. 77: 402-419.

GISI, U. and F.J. SCHWINN. 1976. Phytophthora Newsletter 4: 6-8.

WATERHOUSE, G.M. 1975. Phytophthora Newsletter 3: 5-7.

X. Other aids for the identification of some *Phytophthora*
    species

1. Electrophoresis

Electrophoretic protein patterns have been studied in relation
to taxonomy. In general, the protein patterns obtained corre-
spond to recognized morphological characterization of species.
Electrophoresis thus provide good confirmatory evidence for
identification of species.

Besides protein patterns, intracellular (esterase, alkaline
phosphatase, malic dehydrogenase, and lactic dehydrogenase),
and extracellular (pectic esterase and polygalacturonase) en-
zymes, have also been studied in species of *Phytophthora*. Both
starch gel and polyacrylamide disc gel electrophoresis have
been successfully used to delimit *Phytophthora* species.

CLARE, B.G. and G.A. ZENTMYER. 1966. Phytopathology 56: 1334-1335.

GILL, H.S. and D. POWELL. 1968. Phytopathol. Z. 63: 23-29.

GILL, H.S. and G.A. ZENTMYER. 1978. Phytopathology 68: 163-167.

HALL, R., G.A. ZENTMYER, and D.C. ERWIN. 1969. Phytopathology 59: 770-774.

2. Serology

Immunodiffusion and immunofluorescence tests have been successfully used in studying the relationship between *Phytophthora*
species. It has been shown that effective antigens were obtained with cell-free extracts of mycelium cultured on a glucose-
$(NH_4)_2SO_4$ - fumaric acid medium containing β-sitosterol. (See
page 91 for formula of this medium.)

The stability of the cell-free antigens can be increased by the addition of ethylenediamine tetraacetate to the extraction buffer.

HALSALL, D.M. 1976. J. Gen. Microbiol. 94: 149-158.

MERZ, W.G., R.G. Burrell, and M.E. Gallegly. 1969. Phytopathology 59: 367-370.

3. Thiamine requirement as a means of distinguishing *Pythium* from *Phytophthora*

Most species of *Pythium* can be distinguished from *Phytophthora* on the basis of their requirement for thiamine. Isolates growing in a glucose-asparagine medium without thiamine or with only pyrimidine added, are likely to belong to the genus *Pythium*. Positive identification can then be made by observing zoospore differentiation and release -- *Pythium* spp., characteristically differentiate in a vesicle to the exterior of the sporangium. The following medium is useful for differentiating *Pythium* isolates from those of *Phytophthora*.

a) D-glucose, 10g; L-asparagine, 2 g; KH_2PO_4, 1 g; $MgSO_4 \cdot 7H_2O$, 0.5 g; microelements: Fe^{+3}, 0.2 mg; Zn^{+2}, 0.2 mg; Mn^{+2}, 0.1 mg; Ca^{+2}, 10.0 mg; distilled water, 1000 ml.

b) Remove vitamins present by boiling with activated charcoal (Norit 5 g/liter) for 5 minutes. Filter until clear.

c) Add agar (Difco), 17 g/liter.

d) Most *Pythium* isolates will grow without an exogenous source of thiamine or its moieties (thiazole and pyrimidine), while *Pythium vexans* and *Phytophthora* isolates will only grow when thiamine is added to the medium.

RIDINGS, W.H., M.E. GALLEGLY, and V.G. LILLY. 1969. Phytopathology 59: 737-742.

4. Hymexazol for separation of *Pythium* from *Phytophthora*

Hymexazol (3-hydroxy-5-methylisoxazole)   added   to   selective
antibiotic media (see page 61), will inhibit *Pythium* spp. but
not *Phytophthora*.

MASAGO, H. *et al*. 1977. Phytopathology 67: 425-428.

TSAO, P.H. and S.O. GUY. 1977. Phytopathology 67: 796-801.

5. Pigment production

Pigment production is believed to be a poor taxonomic criterion
due to the variability in the intensity of pigmentation produced by different isolates of the same species. It does however, seem to have value as a confirmatory characteristic once
the species has been identified by conventional morphological
criteria.

When cultured in a casein hydrolysate - tyrosine medium (see
pages 288-289), the following species produced varying degrees
of pigmentation: *P. cactorum*, *P. cambivora*, *P. citricola*, *P.
cryptogea*, *P. drechsleri*, *P. erythroseptica*, *P. infestans*, *P.
megasperma*, *P. nicotianae* var. *parasitica*, and *P. vignae*. It
has been reported that the mean pigment production of A^1 isolates of *P. nicotianae* var. *parasitica* is approximately double
that of the A^2 isolates.

The following species do not produce pigment: *P. cinnamomi*, *P.
citrophthora*, *P. heveae*, *P. phaseoli*, and *P. syringae*.

Note:

Pigment production of all species is inhibited when glucose is
present in the medium.

SHEPHERD, C.J. 1976. Austral. J. Botany 24: 607-617.

TIMMER, L.W. *et al.* 1970. Mycologia 62: 967-977.

5. Malachite green

Malachite green (1 part dye to 1,2,4,8 and 16 ppm of nutrient
solution), has been reported to induce differential growth in
Phytophthora species. Add the dye directly to the nutrient
solution. If dissolved in water first and then added to the
nutrient solution, a precipitate often results at higher concentrations of the dye.

A list of *Phytophthora* species separated by means of differential growth at various concentrations of malachite green can
be found in Leonian, L.H., Amer. J. Botany 17: 671-677 (1930).

Frezzi (1950) and Leonian (1934) have used malachite green to
aid in the construction of keys to *Phytophthora* spp. Grente
(1961) however, found that the pH and N level of the medium
could considerably influence the results obtained. Shepherd
and Pratt (1973) found that malachite green did not sufficiently differentiate the southern and northern ecotypes of *P.
drechsleri* in Australia.

FREZZI, M.J. 1950. Revista Invest. Agric. Argentina 4: 47-133.

GRENTE, J. 1961. Annls. Epiphyt. 12: 25-59.

LEONIAN, L.H. 1934. Bull. Agric. Expt. Sta. W. Va. Univ. 262: 1-36.

SHEPHERD, C.J. and B.H. PRATT. 1973. Austral. J. Biol. Sci. 26: 1095-1107.

7. Cu^{+2} and Pyronin G

An aid to the separation of some *Phytophthora* species and ecotypes of the same species, is the growth response in the presence of Cu^{+2} and pyronin G.

Copper sulfate (33 μg/ml Cu^{+2} ions), and 1 ppm pyronin G has

been reported to successfully distinguished between the 'northern' and 'southern' ecotype of *P. drechsleri* in Australia.

SHEPHERD, C.J. and B.H. PRATT. 1973. Aust. J. Biol. Sci. 26: 1095-1107.

8. Species with close affinities

The following species are closely related to *P. palmivora*: *P.
arecae*, *P. botryosa*, *P. capsici*, *P. colocasiae*, *P. heveae*, *P.
meadii*, *P. mexicana*, *P. nicotianae* var. *nicotianae*, and *P. nicotianae* var. *parasitica*.

With the exception of *P. heveae*, these species have a temperature range similar to *P. palmivora*, i.e. min. 10 C+, opt. 28-
30 C+, max. 35 C+.

P. arecae, *P. citrophthora*, *P. meadii*, *P. mexicana*, and *P. palmivora* are serologically identical and believed to be one species - *P. citrophthora*. Also, *P. drechsleri*, and *P. cryptogea*
are serologically indistinguishable and are believed to be one
species - *P. cryptogea*.

9. Some characteristics separating closely related species

    a) *P. arecae*, and *P. citrophthora* are distinguished from
       P. meadii, *P. mexicana*, and *P. palmivora* on the basis
       of the length-width ratio of the sporangia (1.5:1 or
       more vs. 1.2:1 up to 1.4:1 respectively).

    b) *P. arecae* is separated from *P. citrophthora* based on the
       observations that *P. citrophthora* does not readily pro-
       duce oospores in dual culture, while *P. arecae* produces
       oospores abundantly when paired with the opposite mating
       type.

    c) *P. meadii*, *P. mexicana*, and *P. palmivora*, are separated
       on the bases of the length and shape of the sporangia
       and on the presence or absence of chlamydospores.

       It has been found however, that *P. arecae*, *P. citroph-*

thora, P. meadii, P. mexicana, and *P. palmivora,* produce sporangia with length-width ratios that overlap each other. Also the production of chlamydospores by these five species are extremely variable.

d) *P. drechsleri* and *P. cryptogea* are morphologically similar. Probably the only criterion which may distinguish between these two species is their cardinal growth temperatures - min. ⟨1 C; opt. 22-25°C; max. 31-33°C for *P. cryptogea* and min. 5°C; opt. 28-31°C; max. 36-37.5°C for *P. drechsleri.* However, isolates of *P. drechsleri* are known that do not grow above 35°C. Sporangia sizes vary considerably between these two species (Table 20), making this an unreliable taxonomic character for identification.

P. drechsleri is now considered the same as *P. cryptogea* by most investigators (Bumbieris, M. 1974. Aust. J. Botany 22: 655-660).

0. Other distinguishing characteristics of some *Phytophthora* species

a) *P. nicotianae* var. *parasitica* produces abundant papillate sporangia in both liquid and agar culture. It also has a distinctive slow-growing irregular pattern on agar.

b) *P. cinnamomi* and *P. lateralis* are characterized by distinctive chlamydospores (see Morphology Section).

c) *P. cinnamomi* has hyphal swellings in the shape of large spherical bodies in clusters.

d) *P. cryptogea* and *P. erythroseptica* have small single swellings usually produced in chains.

e) *P. gonapodyides:* The hyphae of this species take on a red-violet color with chloro-zinc iodate.

f) *P. cambivora* oogonia have distinctive irregular bullate protuberances (see Morphology Section).

g) *P. cambivora* and *P. cinnamomi* often have bicellular antheridia.

Table 20: A comparison of sporangia size of various isolates of *Phytophthora*
cryptogea under three nutrient conditions

Isolate	Potato Dextrose Agar (μm)	Pea Broth (μm)	Non-sterile Soil Extract (μm)
Schwinn	24-53 x 18-38	42-74 x 32-51	-
Pethybridge & Lafferty	19-45 x 14-34	38-67 x 27-40	-
Callistephus chinensis	24-56 x 18-38	54-93 x 35-56	-
Solanum melongena	18-67 x 11-31	26-83 x 19-51	-
IMI 21278	-	-	32-60 x 25-30
Native Aust. Shrubs	-	-	35-55 x 28-35
Apricot orchard	-	-	35-55 x 24-40
Apple orchard	-	-	35-55 x 28-44
Pine plantation	-	-	35-65 x 25-38
Cupressus arizonica	-	-	35-77 x 28-44
Peach orchard	-	-	30-50 x 20-30

BUMBIERIS, M. 1974. Austral. J. Botany 22: 655-60. Chitza.ıdis, A. and H.
Kouyeas. 1970. Annls. Inst. Phytopathol. Benaki. 9: 267-274.

h) Some *P. palmivora* isolates often have a high proportion
of double-septate sporangia. See Table 24 for other
characteristics of *P. palmivora* isolates.

i) *P. castaneae* has limoniform sporangia (10-24.5 x 10.0-
37.5 μ); oogonia (19.0-31.0 x 18.0-29.0 μ); oospores
(15-27.5 μ); amphigynous antheridia (7.5-12.5 x 10-
12.5 μ); and chlamydospores (12-19.2 μ). The cardinal
temperatures are min. 9°C, opt. 26-28°C, max. 32°C.
(Katsura, K., 1971).

able 21: Characteristics of *P. citricola* and *P. cactorum*, two species often confused in
      identification

haracteristic	*P. citricola*	*P. cactorum*
porangia apex	Non-protruding with shallow apical thickening.	Hemispherical with deep apical thickening.
porangia shape	Obpyriform, never spherical or ellipsoidal. Often distorted. Sometimes with two widely diverging apices.	Spherical to ellipsoid or obpyriform. Rarely distorted. Sometimes have two papilla, but never widely diverging.
porangia size	21-70 x 15-39 μm. Usually larger than *P. cactorum*.	36-50 x 28-35 μm.
aducity	-	+
porangia stalk	long, non-pedicellate	short, pedicellate
eptal plug	-	+
porangiophores	Long and lax. Branch irregularly, often with swellings at branching.	Compact, fairly regular monochasial sympodium, rarely with branch swellings.
hlamydospores	-	+
ogonia size	21-44 μm, wall 1-1.5 μm thick	25-32 μm, wall 1-1.5 μm thick
ospore size	18-38 μm	20-26 μm
ntheridium	Majority diclinous. Usually arise from a point remote from the oogonial hypha.	Monoclinous. Usually originate from same hypha bearing the oogonia.
rowth	radiating	uniformly fluffy

ef.: WATERHOUSE, G.M. 1957. Trans. Brit. Mycol. Soc. 40: 349-357.

1. Morphotypes of *P. palmivora*

. palmivora isolates present a unique problem in identification.
:solates have been reported to be sufficiently different to be
:ategorized as 'cocoa form', 'Rubber form', and 'Black Pepper
orm'. It has now been found that the portion of stalk attached
:o the sporangium when the latter is detached from the mycelium
.s constant for the different morphotypes. These stalk lengths
ange from short (< 5μm) to extremely long (> 100μm). The black
)epper isolate is further distinguished by having a high pro-
)ortion of double-septate sporangia produced in an umbellate
)r fan-shaped arrangement, compared to the sympodial branching
)f the standard type. The oogonial shapes also differ in the
lifferent morphotypes. Some are spherical, while others taper
.o a funnel-shaped base.

'hese different *P. palmivora* types are presently categorized as

Table 22: Differentiation of *P. citricola* from *P. cinnamomi* on Avocado

Characteristic	*P. citricola*	*P. cinnamomi*
Mycelium	Smooth	Coralloid with abundant hyphal swellings.
☆Growth on Potato dextrose agar	Radiating	Rosette or camellioid.
Oospores	Abundant in single culture.	Usually requires the opposite mating type to form oospores.
Antheridial configuration	Paragynous	Amphigynous
Chlamydospores	-	+
Sporangia shape	Variable, with broad apical thickenings.	Ellipsoid, or ovoid. Non-papillate.
Stalk type	Sympodial	Unbranched

☆Growth on other types of agar media differ, often obscuring the above distinctive growth patterns.

Ref.: ZENTMYER *et al.* 1972-73. Calif. Avocado Soc. Yearbook.

morphological forms (MF1-MF4), the characteristics of which are shown in Table 24.

For detailed discussions of the different forms of *P. palmivora* see:

WATERHOUSE, G.M. 1974. Pages 51-70 *in* Phytophthora disease of cocoa. GREGOR P.H. (ed.) Longman Green & Co. N.Y. 348 p.

TSAO, P.H. and A. TUMMAKATE. 1977. Mycologia 69: 631-637.

SANSOME, E., C.M. BRASIER, and M.J. GRIFFIN. 1975. Nature, London 255: 704-705.

ZENTMYER, G.A., THAWEE KAOSIRI, and GREGORY, IDOSU. 1977. Trans. Brit. Mycol. Soc. 69: 329-333.

Table 23: Summary of some published descriptions of various structures of *Phytophora megasperma* var. *megasperma* and *P. megasperma* var. *sojae*

Hyphae (μm)	Sporangio-phore Lt.	Sporangia Size (μm)	Sporangia Shape.	Chlamydo-spores.	Hyphal Swellings.	Oogonia (μm)	Oospores (μm)	Antheridia (μ)	Cardinal Temp. (C)	Authority
3-8	50-200	35-60 x 25-45	Ovoid	-	-	33-57 Av. 47.4	26-52 Av. 41.4	14-20 x 10-18 65-92% Para-gynous.	2 min. 15-27 opt. 30 max.	Drechsler. J. Wash. Acad. Sci. 1931.
-	-	41.6-56 x 28-40 Av. 49.4x33.6	Obpyriform	-	-	30-50 Av. 40.9	26-45.6 Av. 35.9	-	?min. 25 opt. 30 max.	Tompkins. J. Agr. Res. 1936.
7-8	-	27-63 x 23-38 Av. 31 x 39 40-60 x 24-30	Ovoid to Obpyriform	- -	Terminal Intercalary	23-39 Av. 31	19-37 Av. 25 28-34 Av. 32	7-91% Para-gynous.	8 min. 25 opt. 33 max.	Erwin. Phytopathology. 1954 & 1965.
*2.5-8.0	-	42.5-65 x 31.8-52.5 Av.52.5 x 35.1	Obpyriform	+	-	*29.4-45.1 Av. 36.3	22.8-35.1 Av. 30.7	Paragynous	5 min. 25 opt. 32 max.	Hildebrand. Can. J. Botany. 1959.
1.7-5.1	50-200 Some >400	27-56 x 27-42 Av. 27 x 42	Ovoid to Ellipsoid	+	+	25-41 Av. 35	18-38 20-37	-	5 min. 20-22 opt. 35 max.	Van der Zwet. Phyto-pathology. 1961.
*7-8	-	32-51 x 23-35 Av. 40.5 x 30 119 x 198 24.6-57.8 x 17.2-36.9 Av. 37.0 x 30.2	Ovoid to Obpyriform Oval -	- - - -	Terminal Intercalary - -	*27-54 Av. 35 - 26.2-39.4 Av. 31.5	24-46 Av. 31 25-34 17.0-29.6 Av. 24.3	11-21.5 x 8-16 Av. 15.5 x 13.0 - -	-	Irwin. Aust. J. Exp. Agr. & Husb. 1974. Welty & Busbice. P.D.R. 1976. Leath & Taylor. Pl. Dis. Rept. 1976.
4.2-10	>200	37.4-54.2 x 29-41.6 Av. 46.9 x 33.0	Ovoid	-	Terminal Intercalary	-	-	-	9 min. 27-36 opt. 39 max.	Ribeiro. 1977.

*P. megasperma var. sojae

Table 24: Characteristics of *Phytophthora palmivora* morphotypes

Morphotype	Sporangia	L:B ratio	Caducous	Stalk length (μm)	Chlamy-dospores	Hyphal swell-ings	Chromo-some type	A[1]	A[2]
MF1	35-60 x 20-40 L up to 90	1.67-1.93	Yes	L5	+	−	Small n = 9	+	+
MF2	20-38 x 18-30 L up to 80	1.3-1.4	Yes	5-12	+	−	Small?	+	+
MF3	34-38 x 18-25 L up to 50	1.35-1.47	Yes	5-15	+	?	Small n = 9	+	
MF4☆	34-69 x 19-30	1.7-3.0	Yes	15-250	+/−	−	Large n = 5	+	+
Other†	?	−	No	Remains attached	+	+	?	?	+

☆MF4 is also known as the black pepper form.

†Some isolates are reported to be sterile or neuter.

The large chromosome form (Sansome's 'L' type), has a distinctive nucleus at metaphase with 4-6 chromosomes associated in a ring or chain. It also has a characteristic narrow spindle.

The black pepper form (MF4) of *P. palmivora* from south-east Asia, central America, and Africa has been shown by Tsao (1977 to be sufficiently different from *P. palmivora sensu stricto*, to warrant a new species designation. In addition to their long pedicels, (see Table 24), the black pepper form has a sporangium ontogeny that differs from the typical sympodial formation of *P. palmivora*. The black pepper form sporangia is a fan-shape or umbellate arrangement. The sporangia also have a higher length/breadth ratio than the typical *P. palmivora* type (MF1). The black pepper isolates are also unique in frequently having sporangia that are double septated.

TSAO, P.H. 1977. Abst. 2nd. Intl. Mycological Congress, Tampa, Florida, p. 678.

Section 19:

Texts, review articles, and conferences

I. C.M.I. Distribution Maps of *Phytophthora* Diseases

These loose-leaf maps compiled by the Commonwealth Mycological
Institute, Kew, Surrey, England, give the world distribution
of important plant pathogens and major literature citations.
The maps are periodically up-dated.

The following maps deal with various *Phytophthora* spp.

Map #	*Phytophthora* spp.
35	*citrophthora*
47	*hibernalis*
62	*fragariae*
70	*cambivora*
83	*erythrospetica*
99	*cryptogea*
109	*infestans*
157	*megasperma*
171	*richardiae*
174	*syringae*
201	*phaseoli*
203	*boehmeriae*
204	*porri*
277	*capsici*
280	*cactorum*
281	*drechsleri*
302	*cinnamomi*
428	*heveae*
437	*citricola*
466	*colocasiae*
473	*lateralis*
512	*vignae*

II. C.M.I Descriptions of *Phytophthora* Diseases

These sheets are in loose-leaf form, compiled by the Commonwealth Mycological Institute, Kew, Surrey, England. Each sheet
contains a detailed description of the pathogen, the diseases

caused by the pathogen, physiologic specialization, transmission, geographical distribution, etc.

The following descriptions pertain to *Phytophthora*.

\# 4   *Phytophthora* and *Pythium*

\# 12   *Phytophthora* and *Phytium*

\# 31   *P. hibernalis*

\# 32   *P. syringae*

\# 33   *P. citrophthora*

\# 34   *P. nicotianae* **var.** *nicotianae*

\# 35   *P. nicotianae* **var.** *parasitica*

III. Texts pertaining to *Phytophthora*

1.   Potato Blight Epidemics throughout the World. 1960. Agr.
     Handbook No. 174. U.S. Dept. of Agr., Washington, D.C. 230p

This book contains information on the effectiveness of various
control measures for potato blight caused by *P. infestans*,
throughout the world. It also discusses climatic conditions
conducive to the disease, potato blight forecasting, protective spraying, breeding for resistance, economic effects of
blight, etc.

2.   The Advance of the Fungi. E.C. Large (1962). Dover Publica-
     tions, N.Y. (see pages 13-43).

Contains a chapter describing the destruction wreaked in potato
crops in Ireland in the 1840's by *P. infestans*. A description
of the pathogen is also given. This is a good historical accoun
of *P. infestans* and potato blight.

3.   Foot Rot of *Piper nigrum* L. (*Phytophthora palmivora*). 1963.
     P. Holliday & W.P. Mowat. C.M.I. Phytopathological Paper
     No. 5. Kew, Surrey, England. 62 p.

An extensive treatment of the black pepper disease incited by
P. palmivora. Contains information on early investigations,

isolations, symptoms, pathogenicity, epidemiology, alternate
hosts, control methods, etc.

4.   Disease Resistance in Plants. 1968, J.P. van der Plank.
     Academic Press, N.Y. & London. 206 p.

Discusses data obtained with P. *infestans* as well as other
pathogens in relation to concepts on stabilizing selection,
weak and strong genes, disease resistance (vertical and horizontal resistance) etc.

5.   *Phytophthora* Disease of Plants. 1971. Kiichi Katsura.
     Seibunda Skinkosha Co. Ltd. Japan. 128 p. (In Japanese).

Discusses each *Phytophthora* species separately in relation to
the diseases they cause. Contains descriptions and drawings of
sporangia and oospores of many *Phytophthora* ssp., and photo-
graphs of disease symptoms on host plants.

6.   *Phytophthora* Disease of Cocoa. 1974. P.H. Gregory, (ed.)
     Longman Grp. Ltd. London. 348 p.

Contains chapters on many aspects of P. *palmivora* written by
experts in their respective areas. Topics include a history of
the Black Pod disease, hosts of P. *palmivora*, variation, genetics, distribution of mating types, physiology, methods of testing for resistance, sources of resistance, breeding for resistance, chemical control, etc.

7.   Genetics of Host-Parasite Interaction. 1974. P.R. Day.
     W.H. Freeman & Co., San Francisco. 238 p.

A discussion of the present status of *Phytophthora* genetics is
given on pages 44-51.

8.   Diseases of Cocoa. 1975. C.A. Thorold. Clarendon Press,
     Oxford, England. 423 p.

A comprehensive review of Black Pod disease of cocoa is given
on pages 32-50. It includes information on geographical distribution, economic importance, host range, symptoms, taxonomy,
physiological specialization, spread, predisposing factors, and
control of P. *palmivora*.

9. Monograph on *Phytophthora cinnamomi*. 1979. G.A. Zentmyer.
    Amer. Phytopathol. Soc., Minn. U.S.A.

Contains chapters on all aspects of *P. cinnamomi* from world-
wide distribution and host range to physiology and control.

IV. Review articles on various aspects of *Phytophthora*

1. PETHYBRIDGE, G.H. 1940. The Potato Blight, Yesterday and
    Today. Polytech. Rept. (Royal Cornwall Polytech. Soc. U.K.)
    3: 48-61. An interesting historical survey of the spread
    of potato blight caused by *P. infestans.*

2. HICKMAN, C.J. 1958. *Phytophthora* - Plant Destroyer. Trans.
    Brit. Mycol. Soc. 41: 1-13. An account of the distribution,
    host range, survival, dispersal, and physiological speciali-
    zation of several *Phytophthora* species.

3. ERWIN, D.C., G.A. ZENTMYER, J. GALINDO, and J.S. NIEDER-
    HAUSER. 1963. Variation in the genus *Phytophthora*. Annual
    Review Phytopathol. 1: 375-396.

4. HICKMAN, C.J. and H.H. HO. 1966. Behaviour of zoospores in
    plant pathogenic Phycomycetes. Annual Review Phytopathol.
    4: 195-220.

5. GALLEGLY, M.E. 1968. Genetics of pathogenicity of *Phytoph-
    thora infestans.* Annual Review Phytopathol. 6: 375-396.

6. HENDRIX, J.W. 1970. Sterols in growth and reproduction of
    fungi. Annual Review Phytopathol. 8: 111-130.

    Contains information on the role of sterols in the life
    cycle of the Pythiaceae, particularly *Phytophthora.*

7. TSAO, P.H. 1970. Selective media for isolation of pathogeni
    fungi. Annual Review Phytopathol. 8: 157-186.

    Gives formulae of various antibiotic media suitable for
    selectively isolating *Pythium* and *Phytophthora* directly
    from soil.

8. TUCKER Memorial Symposium on *Phytophthora*. 1969. University
   of Missouri, Columbia, Mo. U.S.A. Published in Phytopathology 60: 1120-1143.

   Contains review articles on development and reproduction
   (Zentmyer & Erwin), genetics (Gallegly), biology of zoospores (Hickman), and taxonomy (Waterhouse).

9. NEWHOOK, F.J., and F.D. PODGER. 1972. The role of *Phytophthora cinnamomi* in Australian and New Zealand forests.
   Annual Review Phytopathol. 10: 299-326.

10. KUC, J. 1972. Phytoalexins. Annual Review Phytopathol. 10:
    207-232. Includes a discussion of phytoalexins from various
    hosts challenged by *Phytophthora*.

11. DICK, M.W. and WIN-TIN. 1973. The development of cytological
    theory in the Oomycetes. Biol. Rev. 48: 133-158.

    Contains a comprehensive review of the cytology and genetics
    of *Phytophthora*. Also puts into historical perspective the
    controversy concerning the ploidy of *Phytophthora*.

12. FOREST FOCUS, Vol. 14, April, 1975. Forest Dept. Govt. of
    Western Australia, Perth. This complete issue is devoted to
    P. cinnamomi and forest destruction.

    It contains a list of shrubs and trees resistant to *P. cin-
    namomi* and also numerous color photographs of various tree
    species destroyed by the pathogen.

13. BARTNICKI-GARCIA, S. and D.E. HEMMES. 1976. Some aspects of
    the form and function of Oomycete spores, pages 594-639 *in*,
    the Fungal Spore: Form and Function (D.J. Weber and W.M.
    Hess, eds.). John Wiley & Sons, Inc.

    Discusses in detail the fine structure of sporangia develop-
    ment and germination, zoospores, cysts, chlamydospores,
    gametangia, and oospores of *Phytophthora*. The article also
    gives an account of various functional aspects of *Phytoph-
    thora* spores, e.g. cell wall glucans, cessation of motility
    of zoospores, aquisition of adhesiveness, etc.

14. ZENTMYER, G.A. 1976. *Phytophthora* - Plant Destroyer. Bioscience 26: 686-688.

A discussion on some aspects of *P. cinnamomi* and avocado root rot.

V. International Conferences on the genus *Phytophthora*

1. Report of the Cocoa Conference, London (1957). Published by the Cocoa, Chocolate and Confectionary Alliance, Ltd. London.

Contains information of research on *P. palmivora* and black pod disease of cocoa.

2. The 2nd and 3rd International Cocoa Research Conferences, held at Bahia, Brazil (1967), and Accra, Ghana (1969), respectively.

Research on *P. palmivora* was reported at both these meetings.

3. 1st International Citrus Symposium. Univ. of Calif., Riverside, Ca. U.S.A. (1968). Proceedings published by Univ. California (Chapman, H.D., ed.), 3 Vols. 1838 p.

Contains information on various aspects of *Phytophthoras* that attack citrus.

4. International Conference on *P. palmivora*. Cacao Research Center, Itabuna, Bahia, Brazil, (Apr. 1971). Proceedings published in Revista Theobroma 1: 55-59.

5. 4th International Cocoa Research Conference. Trinidad and Tobago (1972). Proceedings published in Café, Cacao, Thé 16: 97-110.

6. European Discussion Group on Plant Pathology: Cytology and Genetics of *Phytophthora*. Univ. of Bari, Italy (May 1972). Proceedings published in Bull. Brit. Mycol. Soc. 7: 29-30 (1973).

7. International Seminar on *Phytophthora palmivora*. O.R.S.T.
   O.M. Brazzaville, Popular Republic of the Congo (March
   1973). Proceedings published in Cahiers O.R.S.T.O.M. Biolo-
   gie 20: 1-104 (1973).

8. 2nd International spore Symposium. Brigham Young Univ.,
   Provo, Utah U.S.A. (1974). Published as "The Fungal Spore:
   Form and Function", (D.J. Weber and W.M. Hess, eds.). John
   Wiley & Sons (1976). 895 p.

   Contains an excellent chapter (pages 594-639), by Bartnicki-
   Garcia and Hemmes on form and function of Oomycete spores
   using *Phytophthora* species as examples.

9. Cocoa *Phytophthora* Workshop. Rothampsted Exptl. Sta. England
   (May 1976). Proceedings published in PANS 23: 107-110.

   Includes discussions on the various morphological forms   of
   P. palmivora.

10. International Mycological Congress. Tampa, Florida, U.S.A.
    (August 1977).

    A discussion on the genus *Phytophthora* was held with in-
    vited speakers presenting data on various aspects of the
    genus.

11. Meeting of the Federation of British Plant Pathologists:
    Workshop on the genus *Phytophthora*. Univ. of North Wales,
    Bangor, Gwynedd, U.K. (Sept. 1977).

VI. Other Sources containing information on *Phytophthora*

1. Phytophthora Newsletter. Vol. 1 (Jan. 1973), Vol. 2 (Jan.
   1974), Vol. 3 (May 1975), Vol. 4 (Mar. 1976), Vol. 5 (Feb.
   1977), Vol. 6 (Feb. 1978). Vol. 1-4 published by Ciba-Geigy
   S.A. Switzerland (F.J. Schwinn and D.S. Shaw eds.), Vol. 5
   & 6 published by Dept. of Plant Pathology, Univ. of California, Riverside (G.A. Zentmyer and D.C. Erwin, eds.). These
   newsletters fulfill a useful role in communication between
   Phytophthora workers. Contents include informal articles on

various aspects of *Phytophthora* as well as research findings. Notices of meetings and other announcements pertaining to *Phytophthora* are also included.

2. Avocado Yearbooks - Research on P. *cinnamomi*.

3. Annual reports of the West African Cocoa Research Institute.
   Contains research on P. *palmivora*.

4. American Potato Journal - Research on P. *infestans*.

5. Journal of the Rio Grande Valley Horticultural Society -
   Research on citrus *Phytophthoras*.

Section 20:

Morphology of *Phytophthora* species

The following drawings are not to scale; for dimensions of spore sizes of the various species, refer to the chart on characteristics of *Phytophthora* species.

These illustrations represent redrawings by permission from original papers whenever possible; from G.M. Waterhouse, "The Genus *Phytophthora*," CMI Misc. Publication No. 12 (1956), by permission of the Commonwealth Mycological Institute; from "*Phytophthora* disease of Cocoa" (P.H. Gregory, ed.), by permission of the publishers, Longman Group Ltd. copyright 1968; and from the author's own observations.

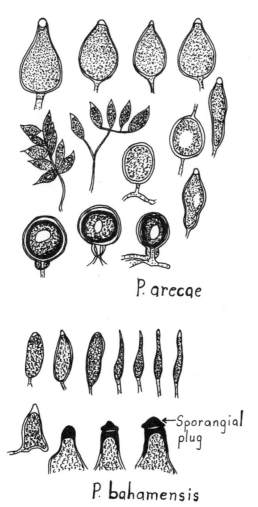

P. arecae

P. bahamensis

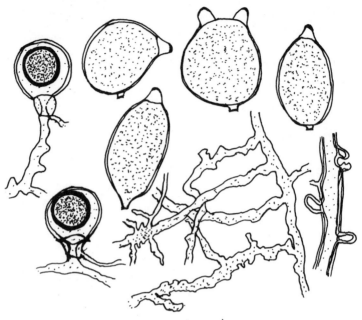

P. boehmeriae

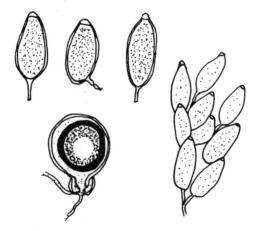

P. botryosa

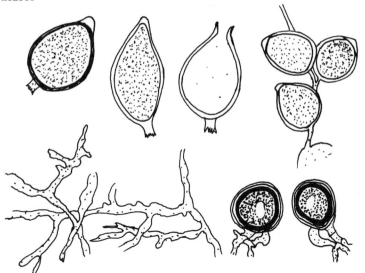

P. cactorum.

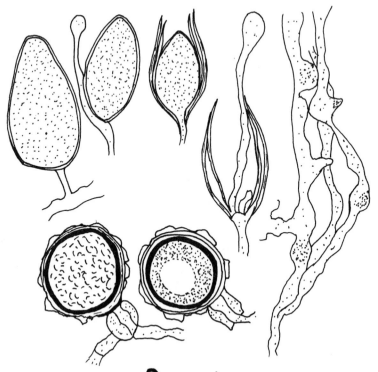

P. cambivora

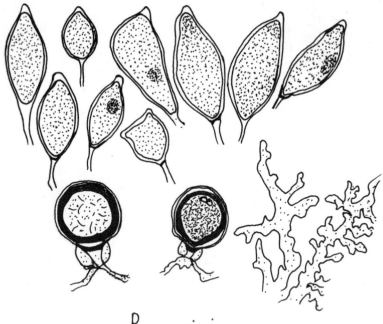

P. capsici

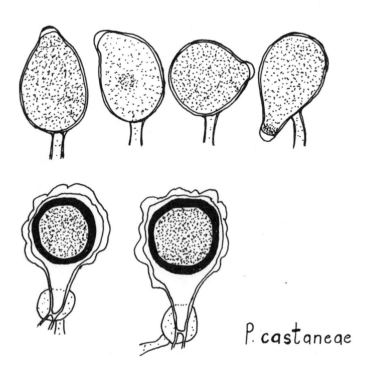

P. castaneae

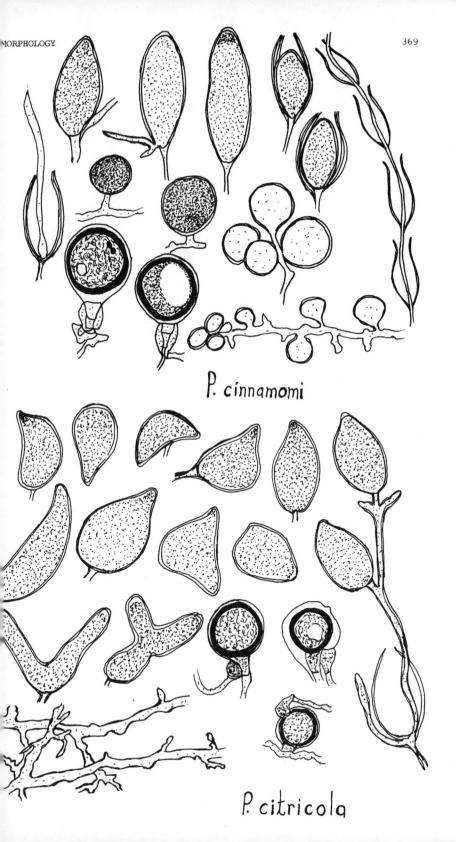

P. cinnamomi

P. citricola

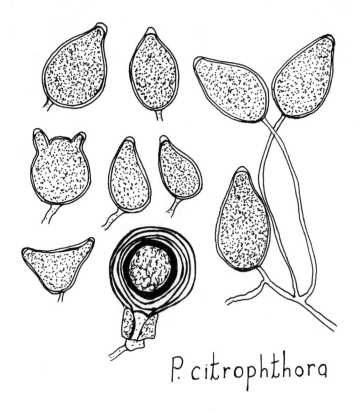

P. citrophthora

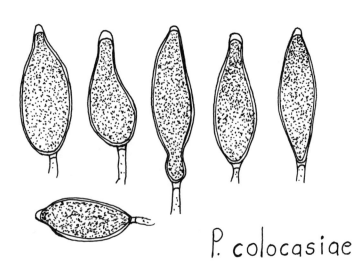

P. colocasiae

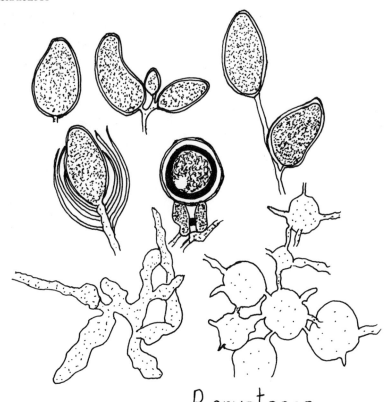

P. cryptogea

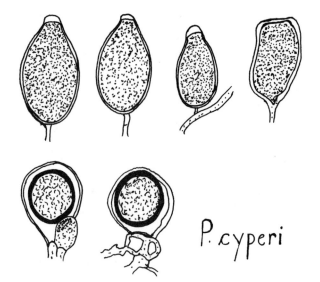

P. cyperi

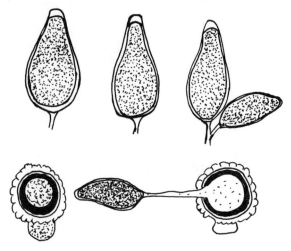

P. cyperi-bulbosi

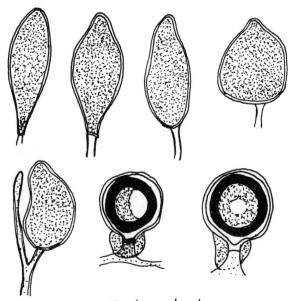

P. drechsleri

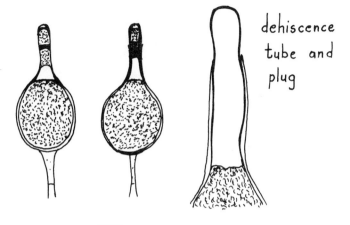

dehiscence
tube and
plug

released dehiscence
plug.

P. epistomium

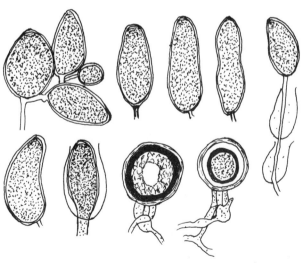

P. erythroseptica

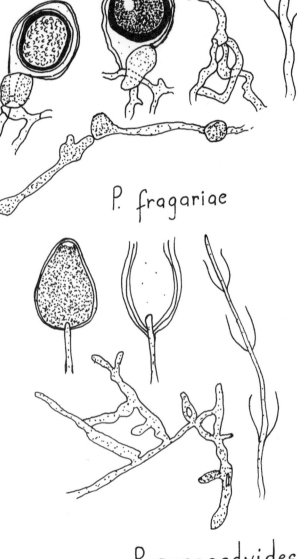

P. fragariae

P. gonapodyides

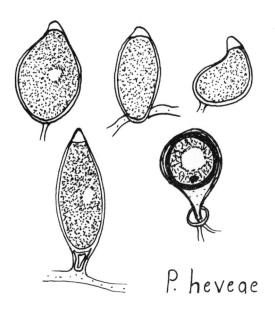

P. heveae

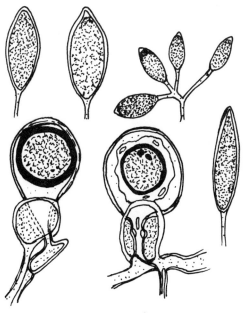

P. hibernalis

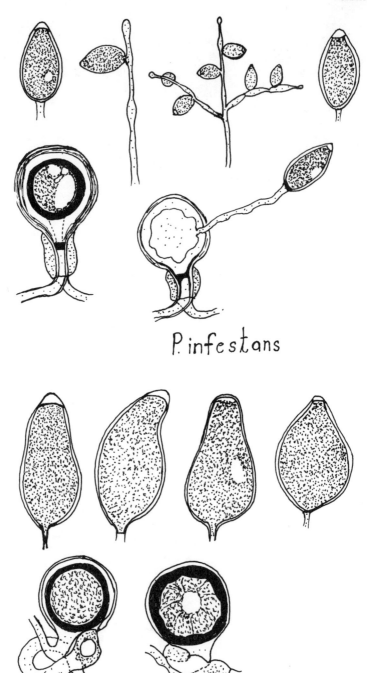

P. infestans

P. inflata

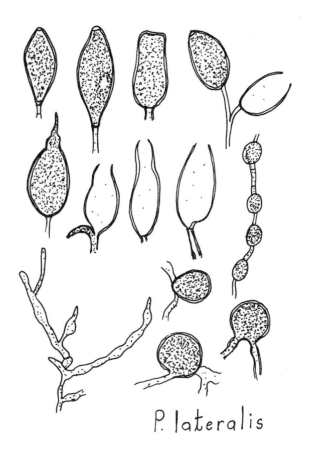

P. lateralis

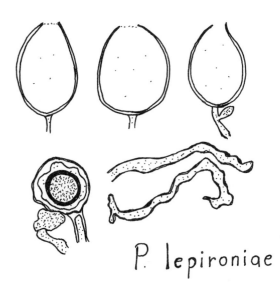

P. lepironiae

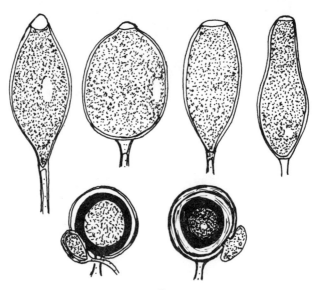

P. macrospora

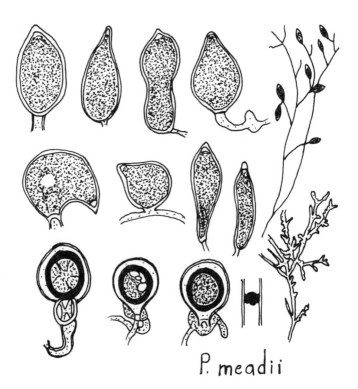

P. meadii

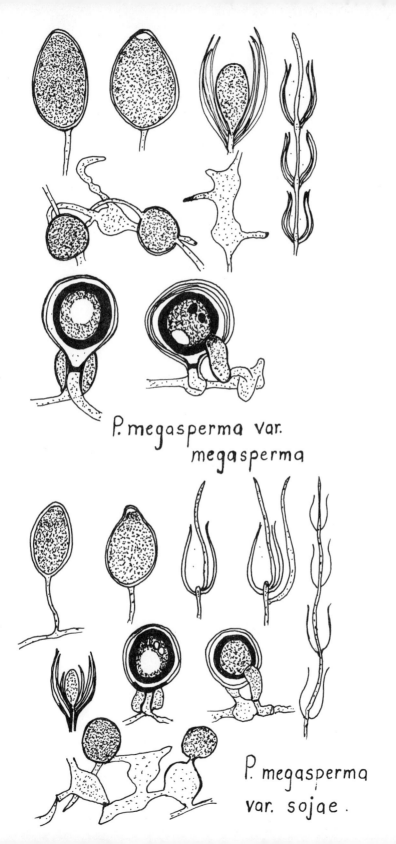

P. megasperma var.
megasperma

P. megasperma
var. sojae.

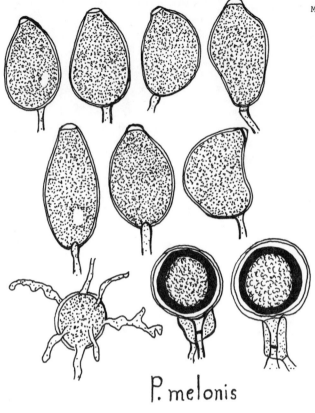

P. melonis

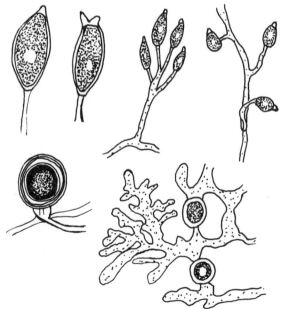

P. mexicana

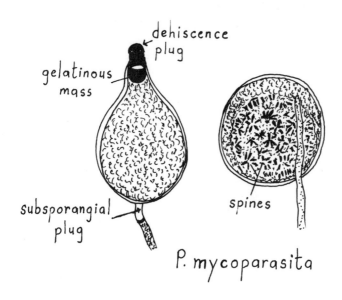

P. mycoparasita

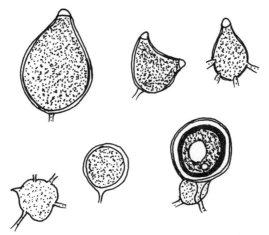

P. nicotianae var. nicotianae

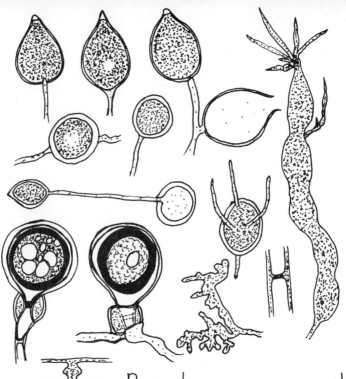

P. nicotianae var. parasitica

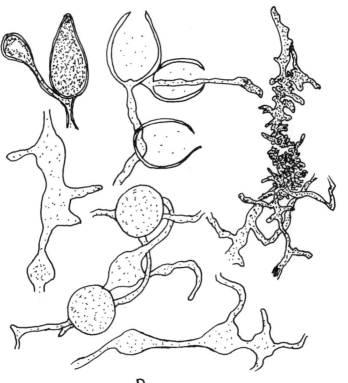

P. oryzae

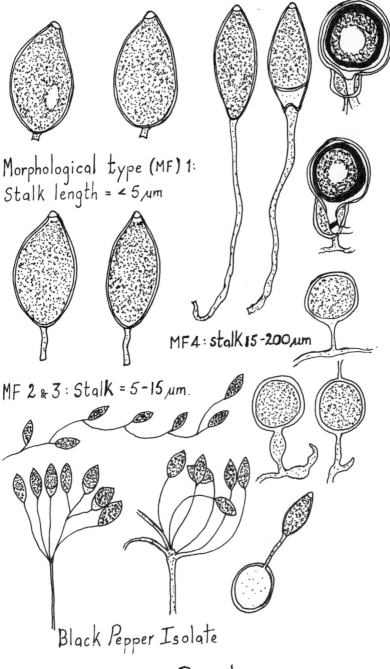

Morphological type (MF) 1:
Stalk length = < 5 μm

MF 2 & 3 : Stalk = 5-15 μm.

MF4 : stalk 15-200 μm

Black Pepper Isolate

P. palmivora

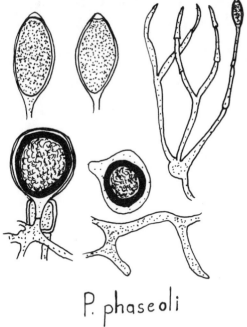

P. phaseoli

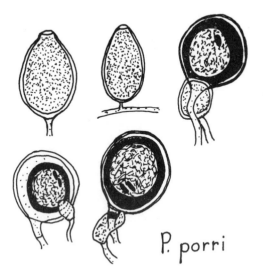

P. porri

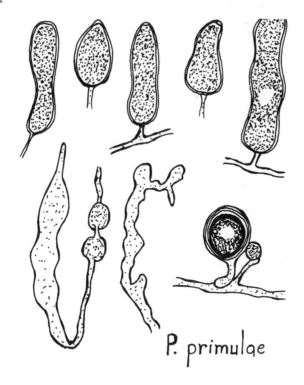

P. primulae

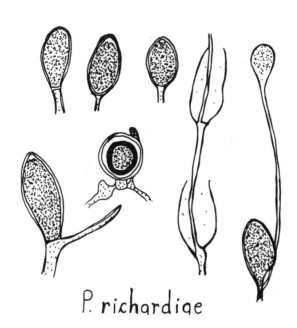

P. richardiae

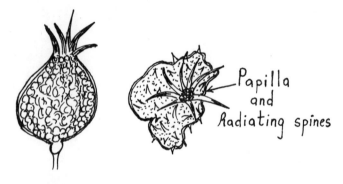

Papilla
and
Radiating spines

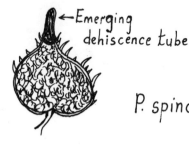

←Emerging
dehiscence tube

P. spinosa var. *lobata*

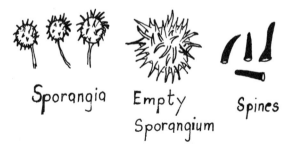

Sporangia   Empty   Spines
              Sporangium

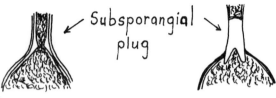

Subsporangial
plug

P. spinosa var. *spinosa*

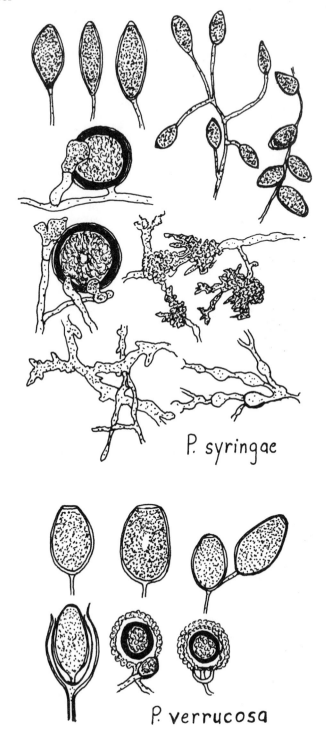

P. syringae

P. verrucosa

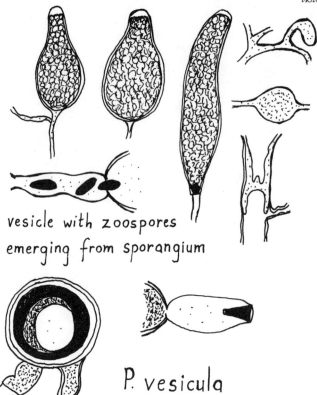

vesicle with zoospores
emerging from sporangium

P. vesicula

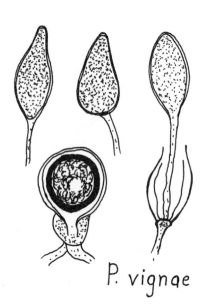

P. vignae

Appendix I

CHARACTERISTICS OF PHYTOPHTHORA SPECIES

Species Name	Hyphal Swellings	Stalk Type	Stalk Length μm	Shape	Length μm	Breadth μm	Papilla	Proliferation	Caducous	Persistent	Chlamydospores	Antheridia Size μm	Paragynous	Amphigynous	Oogonia Size μm	Oospore Size μm	Oospore Wall μm	Homothallic	Heterothallic	Minimum °C	Optimum °C	Maximum °C	Key Group
cactorum	−	Sympodial	<4	Ovoid-Obpyriform	36-50	28-35	+	−	+		+	15x13	+		25-32	20-26	2	+		2	25	30	I
arecae	−	Sympodial	1-6	Ellipsoid-spherical	40-50	35-40	+	−	+		+/−	14x15		+	30-35	26-30	3		+	10-12	27-30	35	II
boehmeriae	−	Sympodial	<3	Ellipsoid-spherical	50	35-40	+	−	+		+/−	14x13		+	27-40	22-26	<2	+		5-6	25	32.5	II
botryosa	−	Branched	5-20	Ovoid-ellipsoid	26-31	14-16	+	−	+		−	14x13		+	25-30	19-24	1-2		+	9	26	32	II
capsici	−	Irreg. branched	10-200	Variable	30-60	25-35	+2	−	+		+/−	17x15		+	30-39	29-38	>1		+	11	28	>35	II
citrophthora	−	Irreg. branched	Long	Variable	45-90	27-60	+2	−	+/−		+	14x14.5		+	26-36	21-28			+	<5	24-28	32	II
heveae	+	Irreg. branched	10	Variable	46-66	29-48	+	−	+		−	9x9		+	25-35	18-23	3		+	11.5	25	32	II
meadii	−	Branched	10-15	Inverse pyriform	33-67	14-28	+	−		+/−	+/−	12x13		+	20-48	16-32	2-4		+	>5	25-30	31-32	II
mexicana	−	Unbranched	Long	Ovoid-fusiform	46-77	24-33	+2	−		+	+	15.5x14		+	30-37.5	28-35	2	+		>10	27.5	>33	II
nicotianae var. nicotianae	+	Sparsely branched	NA	Broadly turbinate	45-70	32-36	+	−		+	+/−	11x10		+	28-30	22-26	1.5-2	+		12	25-30	<35	II
nicotianae var. parasitica	−	Irreg. or Sympodial	2	Pyriform-spherical	38-50	30-40	+	−		+	+	10x12		+	24-31	22-29	2		+	<10	30-32	36.5	II
*palmivora	−	Unbranched or Symp.	4-250	Ovoid-ellipsoid	50-93	31-43	+	−	+		+	15x14		+	25-42	22-39	2		+	11	27-32	35	II
vesicula	+	Sympodial	NA	Ovate-obpyriform	47-192	24-100	+	−	+		+/−	20x9		+	32-60	40-49	2		+	NK	NK	NK	II
carica (fici)	+/−	Unbranched	10-20	Ovoid-ellipsoid	18-25	18-25	+	−		+	+	NK	+	NK	NK	NK	NK		NK	NK	NK	NK	I
melonis ?	?	Sympodial	?	Ovoid	32-75	23-46	+	?	+		+	19x15		+	27-35	25-32	+		+	9	28-32	37	II
citricola	−	Sympodial	NA	Ovoid-obclavate	21-70	15-39	+/−	?	−	+	−	12.5x9		+	21-44	18-38	1-2	+		3	25-28	31	III
cyperi	−	Sympodial	4	Ovoid-pyriform	52-130	30-44	+/−	−	−	+	−	20x12		+	37-48	35-45	3	+		NK	NK	NK	III
cyperi-bulbosi	?	Sympodial	?	Obpyriform	20-67	23-34	+/−	−	−	+	−	16x12		+	31-53	25-42	1-2	+		NK	NK	NK	III
inflata	+	Unbranched	NA	Limoniform-elongate	20-67	15-32	+/−	−	−	+	+	?		+	30-43	26-39	−	+		NK	25-30	<35	III
lepironiae	+	Unbranched	5-20	Ovoid	40-64	36-52	+/−	−	−	+	−	23x15		+	40-52	30-38	3-4	+		NK	NK	NK	III
macrospora	+	Sympodial	14	Ovoid-ellipsoid	85-112	50-65	+/−	−	−	+	−	27x15		+	50-100	45-94	3-6	+		18-19	28	32	III
porri	+	Unbranched	NA	Obpyriform-ellipsoid	35-80	29-50	+/−	−	−	+	−	12x12		+	36-45	33-42	4.5	+		<5	25	33-35	III
primulae	?	Unbranched	10-500	Obpyriform-ellipsoid	33-165	23-52	+/−	−	−	+	−	15x15		+	23-43	17-33	3	+		<5	NK	NK	III
syringae	+	Sympodial	NA	Ovoid-obpyriform	57-75	36-42	+/−	−	−	+	+	10x7		+	33-46	31-43	2	+		<5	20	23	III
colocasiae	−	Irreg. branched	3.5-10	Ellipsoid-fusiform	45-70	23-28	+/−	−	+		+	12x11		+	26-35	23-32	2.5		+	>10	27-30	>35	IV
hibernalis	−	Irreg. sympodial	55	Ovoid-fusiform	36-56	19-28	+/−	−		+	+/−	12x10		+	35-56	28-32	3		+	<5	20	25	IV

Species	(+/−)	Branching	Shape	50-200	32-51	23-35	(−)	(+)	(+)	(−)	15×13	(+)	27-54	24-46	1-3	(+)	5	25	33-35	(VI)
megasperma var. sojae	+	Unbranched	Ovoid-obpyriform	50-200				+	+	−	Variable	+	67-82	49-64	?	+	NK	NK	NK	VI
quininea	+	Unbranched	Ovoid-obpyriform	75	46-68	25-38	−	+	+	+	16×13	+	23-47	17-31	3	+	NK	NK	NK	VI
verrucosa	−	Unbranched	Ovoid-obpyriform	15-250	41-56	29-36	−	+	+	?	16×13	+				+	NK	NK	NK	VI
cambivora	+	Unbranched	Ovoid	NA	55-65	40-45	−	+	+	+	25×35	+	43-62	32-38	2	+	2	22-24	32	VI
cinnamomi	+	Unbranched	Ovoid-ellipsoid	NA	27-114	20-71	−	+	+	+	22×17	+	21-58	19-54	3.5	+	5-16	20-30	30-36	VI
cryptogea	+	Sympodial	Ovoid	NA	37-55	23-30	−	+	+	+	10×10	+	30-38	27-35	3	+	<1	22-25	31-33	VI
drechsleri	+	Sympodial	Very variable	NA	36-70	26-40	−	+	+	+	14×13	+	36-53	33-50	2.5	+	5	28-31	36-37	VI
erythroseptica	+	Sympodial	Ellipsoid-obpyriform	NA	43-69	26-47	−	+	+	NK	14×13	+	30-46	27-43	2.5	+	<2.5	27.5	34	VI
gonapodyides	+	Unbranched	Obpyriform	NA	42-70	20-32	−	+	+	NK	NK	NK	NK	NK	3	NK	NK	NK	<30	VI
lateralis	+	Sympodial	Ovoid-obpyriform	NA	26-60	12-20	−	+	+	+	12×18	+	NK	NK	−	+	<2	20	35	VI
oryzae	+	Branched	Ovoid-ellipsoid	NA	41-84	20-40	−	+	+	+	?	+	33-50	28-46	4	+	9-11	26-28	32	VI
richardiae	+	Unbranched	Very variable	NA	40-60	20-40	−	+	+	?	18×14	+	24-38	20-33	3	+	3-4	24	34.5	VI
vignae	+	Unbranched	Ovoid-obpyriform	NA	48-72	27-54	+	+	+	−	16×15	+	32-46	26-32	3	+	10-12	28-30	35	VI
bahamensis	−	Unbranched	Bursiform-multilobe	NA	26-119	19-43	Plug	−	+	NK	NK	NK	NK	NK	−	NK	?	28-33	35	−
epistomium	−	Unbranched	Lageniform-obpyri.	NA	43-184	56-107	Plug	−	+	NK	NK	NK	NK	NK	−	NK	?	28-30	35	−
mycoparasitica	−	Unbranched	Obnapiform	NA	26-131	14-111	Plug	−	+	NK	NK	NK	NK	NK	−	NK	?	?	39	−
spinosa var. spinosa	−	Unbranched	Ovate-globose	NA	?	60-107	+	−	+	NK	NK	NK	NK	NK	−	NK	?	28-37	?	−
spinosa var. lobata	−	Unbranched	Obpyriform-auriculate	NA	51-75	56-150	+	−	+	NK	NK	NK	NK	NK	−	NK	?	28-33	37	−

*For Morphotypes of P. palmivora see text: NK = Not known: NA = Not applicable: Note: Some Heterothallic species can produce oospores in single culture.

© Olaf K. Ribeiro 1977

Illustrated by Wendy Reid - November 1977, Riverside, California.

Appendix II

Diagrammatic representation of the life

cycle of *Phytophthora*

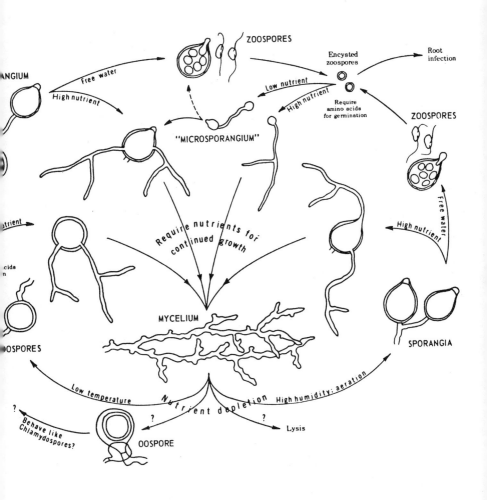

Saprophytic phase of the life cycle of *Phytophthora parasitica* showing the
effects of certain physico-chemical factors discussed in the present paper
on the saprophytic behavior of the fungus in soil as well as in vitro. By
permission of the author (P.H. TSAO) from Proc. First Intl. Citrus Symposium
(1969).

Appendix III

Spectra of some lamps commonly used in

Phytophthora research

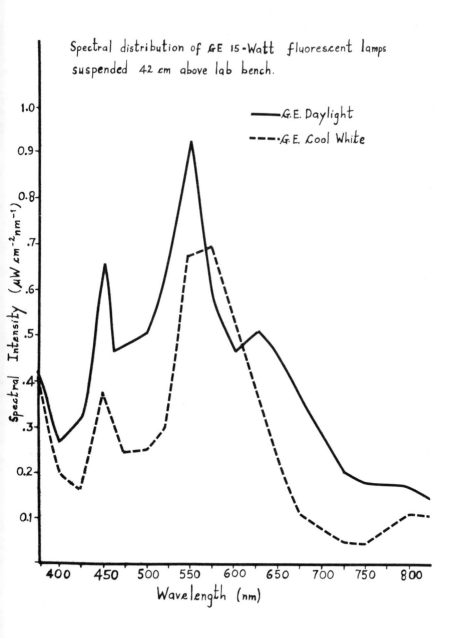

Spectral distribution of GE 15-Watt fluorescent lamps suspended 42 cm above lab bench.

——— G.E. Daylight

- - - - G.E. Cool White

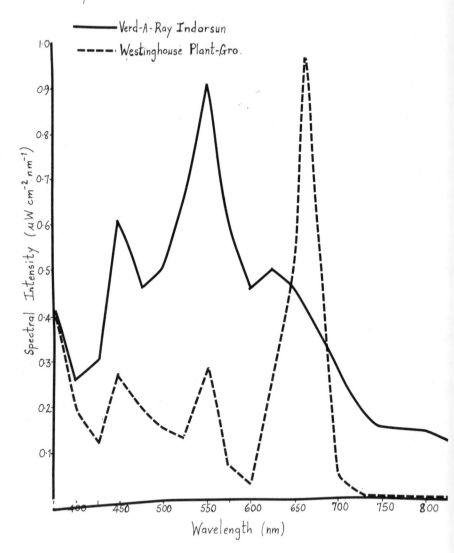

Spectral distribution of two types of fluorescent lamps (15w) suspended 42 cm above lab bench.

——— Verd-A-Ray Indorsun
- - - - Westinghouse Plant-Gro.

Spectral Intensity ($\mu W\,cm^{-2}\,nm^{-1}$)

Wavelength (nm)

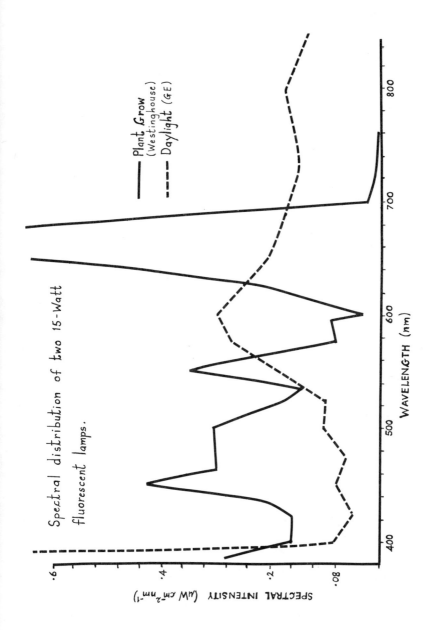

Spectral distribution of two 15-Watt fluorescent lamps.

Plant *Grow* (Westinghouse)
Daylight (GE)

SPECTRAL INTENSITY (μW cm⁻² nm⁻¹)

.6

.4

.2

.08

WAVELENGTH (nm)

400   500   600   700   800

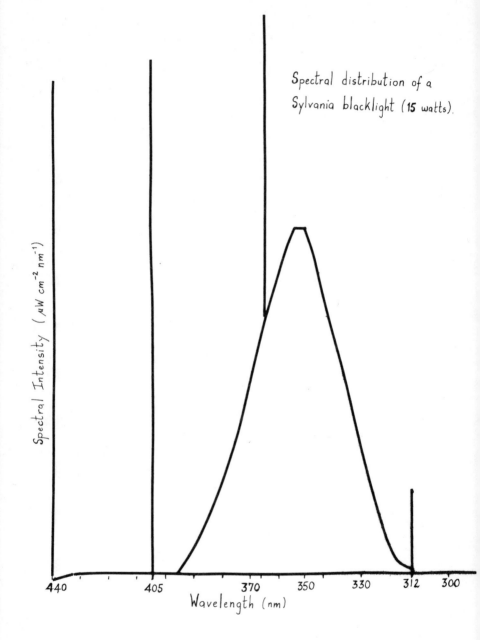

Spectral distribution of a
Sylvania blacklight (15 watts).

Spectral Intensity (μW cm^{-2} nm^{-1})

Wavelength (nm)

440   405   370   350   330   312   300

Index